# Plastic Bangles

Lyn Tortoriello and Deborah Lyons

Photography by Lyn Tortoriello

4880 Lower Valley Road, Atglen, PA 19310 USA

Cataloging in Publication Data:

Tortoriello, Lyn.
Plastic bangles / by Lyn Tortoriello and Deborah Lyons.
p. cm.
ISBN 0-7643-2195-1 (hardcover)
1. Plastic jewelry—United States—History—20th century.
2. Bracelets—United States—History—20th century. I. Lyons, Deborah. II. Title.
NK4890.P55T67 2005
688'.2—dc22
2004026702



Designed by John P. Cheek
Cover design by Bruce Waters
Type set in Arial

ISBN: 0-7643-2195-1
Printed in China
1 2 3 4

Published by Schiffer Publishing Ltd.
4880 Lower Valley Road
Atglen, PA 19310
Phone: (610) 593-1777; Fax: (610) 593-2002
E-mail: Info@schifferbooks.com

For the largest selection of fine reference books on this and related subjects, please visit our web site at
**www.schifferbooks.com**
We are always looking for people to write books on new and related subjects. If you have an idea for a book please contact us at the above address.

This book may be purchased from the publisher.
Include $3.95 for shipping.
Please try your bookstore first.
You may write for a free catalog.

In Europe, Schiffer books are distributed by
Bushwood Books
6 Marksbury Ave.
Kew Gardens
Surrey TW9 4JF England
Phone: 44 (0) 20 8392-8585;
Fax: 44 (0) 20 8392-9876
E-mail: info@bushwoodbooks.co.uk
Free postage in the U.K., Europe; air mail at cost.

# CONTENTS

# Acknowledgments

We have been fortunate in having the best of editors and publishers – Nancy and Peter Schiffer – whose hospitality and helpfulness are unmatched. Bruce Waters of Schiffer Publishing patiently initiated us into the mysteries of the medium format camera and designed the elegant cover. Thanks also to the rest of the staff at Schiffer Publishing for their help and encouragement.

The majority of the pieces shown are from our own collections, but we gratefully acknowledge the generosity of Lori Kizer of the Rhinestone Airplane (www.rhinestoneairplane.com), who lent us pieces by Judith Evans and Alexis Bittar. Bill Ruffin of the Bangle Barn (thebanglebarn@aol.com) and Lisa Sachs of Bakelite Square at www.Rubylane.com helped us with loans of Asian fakelite. Joan Young at Trojan Antiques, Alexandria, Virginia (bonnetstobustles@msn.com) and Harry Flores of the Bethesda, Maryland, Farmers Market each lent us a variety of pieces. Ann Barbour of Ann's French Twist (Ebay store) lent us the Rafia & Bossa bangles. Véronique Cassel of Iris Bleu Boutique (Ebay store) provided us with invaluable information about French plastics and designers, and has been a source of inspiring French vintage pieces. Frank Kelly and Normand Thibaudeau answered our many questions about the history and manufacture of Best Plastics bangles. Leonore Moog kindly provided a piece from her collection. Mary Fletcher's friendship and generosity deserve special mention. Collectors and dealers too numerous to mention have provided us with information along the way, putting up with our sometimes strange questions about labels, materials, and other arcana. Friends and colleagues have tolerated and sometimes even shared our enthusiasm for vintage plastic. The employees of Highfield House in Baltimore graciously handled the hundreds of packages that arrived while we were gathering bangles to photograph.

We hope to hear from our readers. All comments, questions, and corrections may be addressed to toriello@msn.com and basilissa@yahoo.com or you may write to us care of Schiffer Publishing.

Chapter 1

# Introduction: Bewitched, Bangled, and Bewildered

What is it about the bangle that is so appealing? From at least the third millennium B.C.E., human beings have been making and adorning themselves with circlets of different materials, ranging from bone to glass to gold. The excavated graves of nearly every early civilization have yielded up bangles among the jewels left to accompany the departed, and pictorial art shows us that they also adorned the living. In some traditional cultures the bangles a woman wears make up a significant portion of the family's stored wealth. A bangle embodies the infinity and perfection of the circle, while providing a showcase for the skill of the artisan who made it. Both as adornment and display object, the bangle is uniquely satisfying.

In twentieth-century America, the plastic bangle had periods of popularity alternating with obscurity, reflecting the ups and downs of our history. During the depression, plastic jewelry allowed the wearer a bit of light-heartedness in difficult times, while during World War II, it replaced metals needed for the war effort. Its history is closely tied to the history of technology. As new plastics came on the scene, they were commandeered for the making of bangles. We may smile at the innocence and optimism of the mid-20th century when American dreams seemed so obtainable. Colorful, fun, whimsical bangles poured out of factories by the thousands. Yet, pride of workmanship is still alive in the memories of those who made them. And rightly so; years later we can admire a bangle that still looks as good as the day it was made. Whether it was decorated by the artisan, a machine, or a combination of both, we can see the critical role played by the human hand. As we revel in these delightful pieces, we salute the chemist, the carver, the machinist, and the artist. Their efforts have not been lost.

Popular off and on again from the 1950s through the1980s, when many of our examples were made, bangles almost disappeared for several decades, only to experience a vigorous revival in the early years of the new millennium. As we write this book in the summer of 2004, plastic bangles are appearing in stores and fashion magazines. Some are self-conscious evocations of the bangles of thirty or forty years ago, while others are the work of contemporary artists. Perhaps not coincidently, collectors' interest in Lucite and other vintage bangles has risen dramatically in the last couple of years.

When we first started collecting plastic bangles, we found ourselves in a quandary. Bewildered by the variety of plastics on the market, we soon realized that there was no comprehensive guide we could consult. There were so many different kinds and so many conflicting opinions about their identity. We both began our collections with Bakelite, but over time we were increasingly attracted to more recent plastics, which were often more colorful and affordable. As we stepped up our collecting, bangles continued to hold pride of place in our affections, although at times we strayed from

Mountain of Bangles.

the true path to include matching earrings, pins, clips, rings, and necklaces, as well as other kinds of bracelets. We set about learning as much as we could about our growing collections. This book is the result of our efforts to sort out this complicated but rewarding new collecting field. In it we share with you what we have learned about the materials, techniques, designs, and origins of these amusing and delightful pieces of jewelry.

Once we decided to write this book, we began amassing what turned out to be thousands of bangles and other bracelets. Fearful of the "collector detector" (that nagging inner voice that reminds you what your collection is missing) we tried to gather every single type in every size and color. If anyone was bangled, we were! Yet, even this overwhelming quantity represents only a fraction, albeit a large fraction, of the types that were produced. We have tried to steer a path between completeness and insanity –a balancing act any true collector will understand. We were truly bewitched.

Our emphasis is on pieces produced from the 1950s to the 1980s, most of which are made of Lucite or similar plastics. Although the bulk of the bangles shown here were made in the U.S. or for the North American market, we include pieces from other countries, as well as the work of foreign designers. In order to give a full picture of plastic bangles of the twentieth century, we have included older plastics, such as celluloid, galalith, and Bakelite. We end with a survey of newer pieces of two kinds. The first group we include to alert collectors to recent imitations of older pieces, most rather cheaply made in Asia. The second group consists of contemporary designers we admire and whose work we consider highly collectible now and for the future.

No author is infallible, and we fully expect, despite our best efforts, to be guilty of omissions and errors. For this reason, we regard this book as an on-going process and welcome all additional information that our readers may have to offer. With luck and your assistance, we hope to be able to acknowledge you in a future edition. May you enjoy this book and find as much pleasure in collecting as we have!

## In Defense of Plastic

As synthetic plastics came into use, the word *plastic* acquired a bad connotation in some quarters, suggesting artificiality or inauthenticity. (For discussion of the cultural meanings of "plastic," see Meikle, *American Plastic*, pp.6-9.) In the 1960s, uneasiness over artificial substitutes for "natural" items turned the word into an insult. "Plastic" people were especially to be avoided. But the word *plastic* comes from the Greek verb *plassein*, to form, mold, or shape, and means "capable of being molded." When an ancient Greek potter made a vase, this was the verb he would use to describe his activity. Today, plastic objects from the mid-twentieth century are being recognized as design classics. While we may still feel some unease about our society's reliance on the petrochemicals from which many plastics are derived, it is time to redeem the word.

In our experience, people can become offended if told that their piece is not Bakelite but "just plastic." So for the record, Bakelite is plastic, celluloid is plastic, Lucite is plastic. Plastic is good! We like plastic! In fact, we have become plastic fanatics. Beware – it could happen to you too.

## A Note on Values

As with all collectibles, prices vary greatly. The pieces in this book range from durable modern plastics manufactured in large quantities, to Bakelite pieces hand-carved in small numbers, to perishable celluloid or galalith. Rarity and condition, as well as popularity with collectors, play a role in determining price. Trends in collecting are cyclical. At the present time, Bakelite, while still popular, is not commanding the eye-popping prices of a few years ago, while fine celluloid pieces are favored by high-end collectors. Meanwhile, Lucite is coming into its own for the first time as a highly sought-after collectible. We believe that good Lucite pieces will continue to appreciate for quite awhile.

All of our prices assume a piece in good to excellent condition with no damage and no repairs. We give values as a range, with the lower number representing a "snap-it-up" price, and the higher, a price at which the item commonly sells on the open market. This higher number represents a high but not extortionate retail price. Occasionally something will sell for double or triple our high price, but since these exceptions are usually the result of inaccurate descriptions or cut-throat bidding between two impassioned collectors, we do not take account of them. We have both paid more than our high price for items we just *had* to have for our personal collections. This is perfectly all right – you will do it too. Think of the prices listed as providing a rough guide. They may not always answer the question, "Is it worth $2 or $12?" but should help when the question is, "Am I crazy to pay $200?"

A group of Lucite spacer bangles in a clear plastic box with a loop for hanging, 1960s. Price marked is $1.00.

Chapter 2

# A Very Brief History of Plastics

At the beginning of the twentieth century, materials used for jewelry included natural plastics such as amber, horn, and rubber, as well as a few partly man-made plastics like celluloid and casein. "Plastics" indicate materials that change their character under the influence of heat. These fall into two basic types: **thermoplastic** and **thermosetting** plastic. Despite the apparent similarity of the names they are very different. The majority of plastics are thermoplastic, which means that when heated they become malleable. These include celluloid, Lucite, and other modern plastics like polyethylene. Thermosetting plastics are quite the reverse, as once the hot liquid has cooled, hardened, and cured, it can no longer be melted down and re-formed. Bakelite is the most notable of the thermosets. Understanding the difference between these types is one of the keys to plastics identification. A number of excellent books provide the details of this history (see "Suggestions for Further Reading"), so we will limit ourselves to the kinds of plastics most often used for bangles.

**Celluloid**, patented in 1868, is a tradename for cellulose nitrate, which is made from cotton fibers and nitric acid. It is a thermoplastic, and when hot can be twisted and bent as desired. Even when cool it is more or less flexible depending on its thickness. In the nineteenth and early twentieth centuries, celluloid had an amazing range of applications from film to shirt collars, and was widely used in jewelry-making. It was frequently stamped or molded into bracelets imitating ivory, many made in Japan. It was also decorated with carving, with rhinestones (in so-called "sparkle bracelets"), or hand-painted. Sometimes these techniques were combined.

A group of celluloid floral bangles and a celluloid chrysanthemum cuff. The two white hinge bracelets on the right are made of an artificial compound intended to imitate celluloid. Celluloid bangles, $25-75, others $15-20.

An elegant black and white set. Celluloid bangle with matching earrings and a jabot pin. Bangle, $65; earrings $25-35; pin, $35.

Celluloid bracelets in many different styles and techniques.

Celluloid buckle-type bracelets. Left, with rhinestones and faux pearls; right, molded floral, $45-50.

Three celluloid bracelets, each with one end that pulls through or wraps around the other end. Black bracelet, $75, gold, $55, cream, $40.

Stack of thin bangles, probably celluloid. Set of five layered multicolored bangles, $35-45; pair of black with mother-of-pearl, $25-30.

Two celluloid "guitar pick" stretchy bracelets, $35-45.

Impressive and unusual celluloid cuff, black layered over yellowed clear, painted with rhinestones, $75-95.

Celluloid deco sparkle bracelets, $35-75; with circles, $500 and up.

The earliest celluloid was highly inflammable and resulted in terrible accidents in factories and cinemas. Over time, a more stable version, **cellulose acetate**, was developed, which is still in limited use today, most notably by the French designer **Lea Stein**. Meanwhile celluloid ceased to be produced in the 1950s. These plastics can be identified by their lightness and flexibility, and in the case of celluloid, by smell (see "Identifying Plastics" below). When molded they often display an intricacy and consistency of detail that betrays their mechanical production.

Group of cellulose acetate bangles, bypass bangles, and cuffs.

Cellulose acetate bracelets. Blue and tortoise wrap bangles, $35-40; cuff, $25; bypass, $35.

Lea Stein layered bangle, $300. Courtesy Nancy Schiffer.

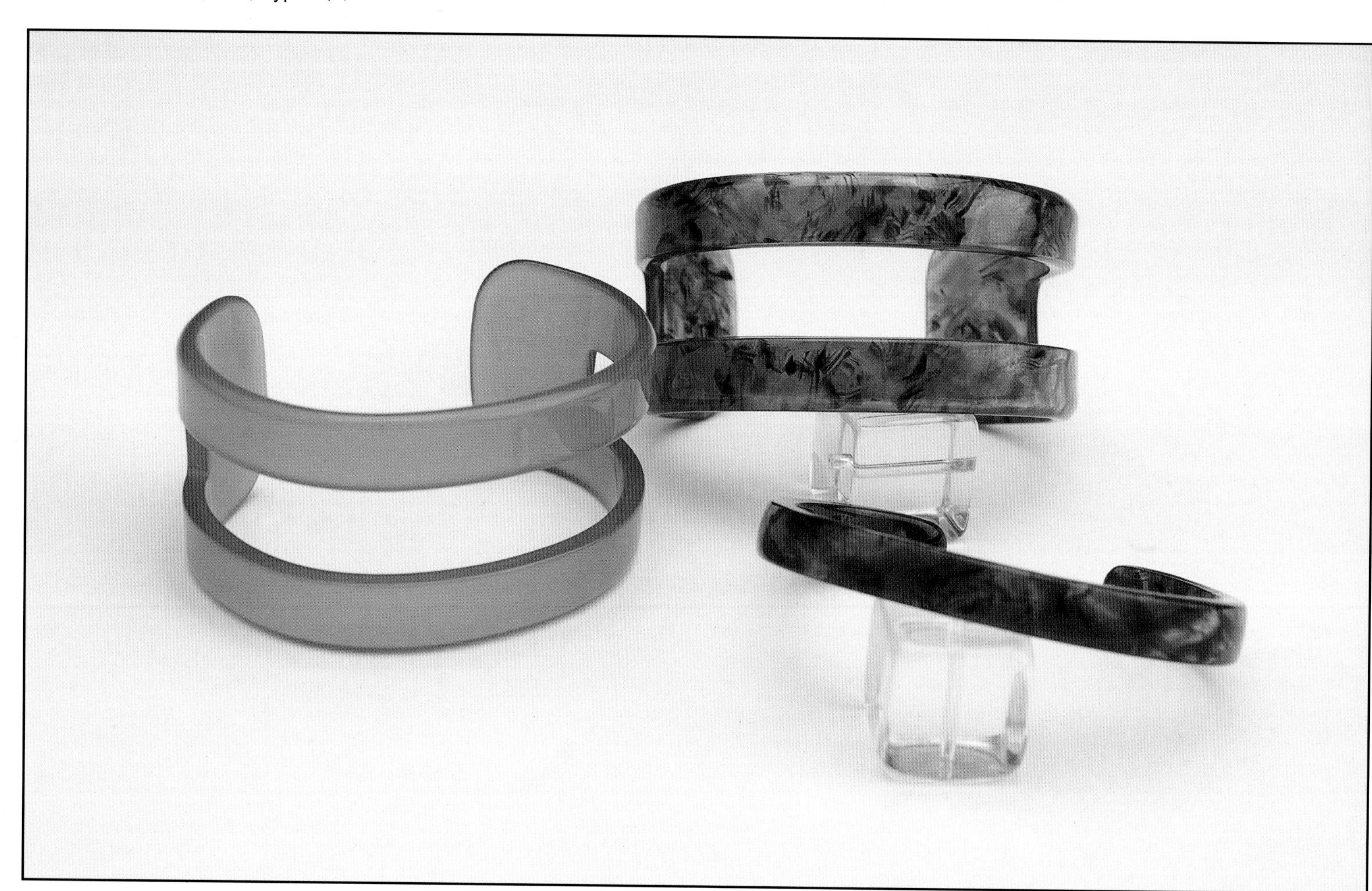

Three cellulose acetate cuffs, recently made by a company that also makes eyeglass frames, $15-35.

**Galalith** (from the Greek *gala*, milk, and *lithos*, stone) is a milk-based plastic, produced in France, which was used to make jewelry, among other things. (Its American counterpart, casein, is much less important in jewelry production). Originally white in color, it can be dyed, and can achieve the depth of color found in Bakelite. Earlier galalith has the depth and glossiness of Bakelite, while later galalith is softer in color. The production of this plastic peaked in the 1920s and 1930s and ceased almost completely by the 1980s. Its fashion heyday was the 1930s and 1940s, at the height of Bakelite's popularity. In France, the two plastics were sometimes assembled or even layered to make up one piece. Some old stock is still available and is being used by a few contemporary French designers of pins. One company in Italy still produces galalith, but only in sheets, in black and white. Vintage galalith jewelry is rare and highly collectible. **We do not believe that any bangles are currently being made of galalith.**

Three galalith bangles with shaded color wash. Art nouveau style purple and pink bangles with openwork, $95 and up; pink floral carved, $35.

Galalith is sometimes called "**French Bakelite**," a misleading term that has caused untold confusion. We call for the abolition of this term, since it obscures the fact that much true – and very beautiful — Bakelite was made in France. Bakelite made in France passes the same tests as American Bakelite, with only slight differences. (See "Testing.") Recently, unscrupulous or ill-informed dealers have been using this term to "dress up" newly made and undistinguished Lucite and hard plastic pieces. Adding to the confusion, some people claim that galalith or "French Bakelite" is Bakelite "without the chemicals," which is technically impossible. All artificial plastics are made with chemicals, although some modern formulations omit the more toxic substances that were once used. Similarly meaningless is "low (or no) odor" Bakelite – another attempt to pass off modern substances as vintage.

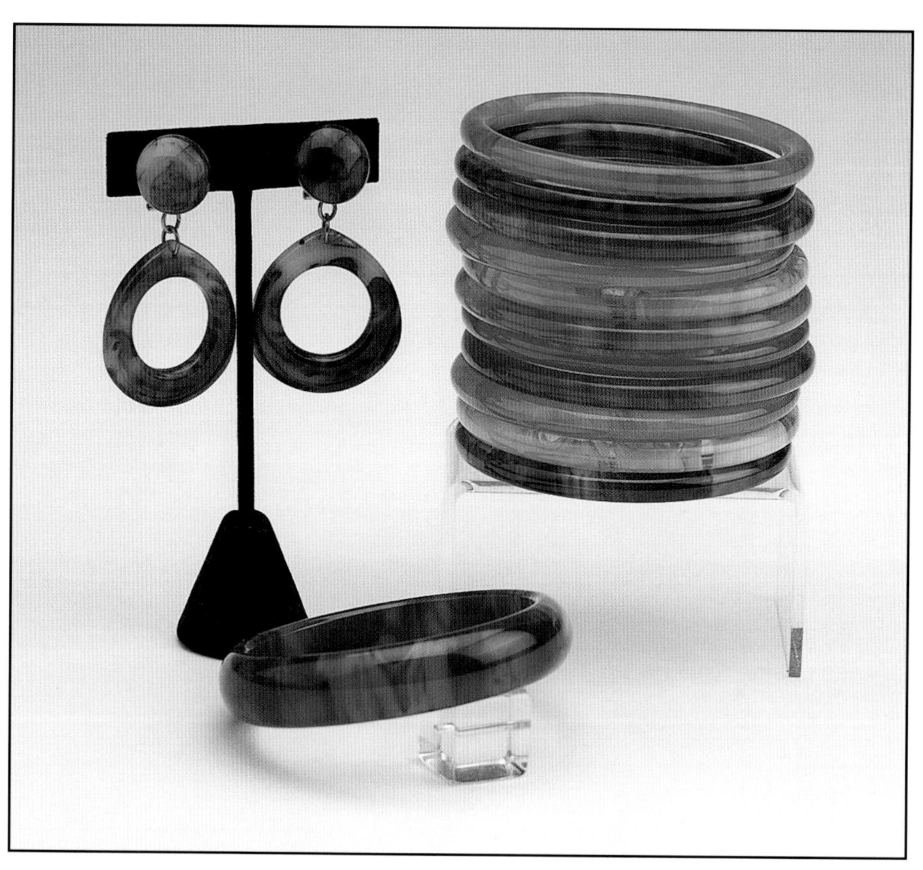

Typical colors of Bakelite made in France. The earrings are instantly recognizable by the loop clip-back typical of French manufacture. Spacers, $20; bangle, $35-40; earrings $30–35.

A stack of honey and tortoise Bakelite bangles made and purchased in France, $40-50 each.

Marbled pink Bakelite from France. Bangles, $35-60; necklace, $45-55; earrings, $20, 30.

**Bakelite** is the first wholly artificial plastic, and was invented in 1917 by the Belgian chemist Leo Baekeland. Its chemical name is phenol formaldehyde resin and it is a thermoset plastic. At first it was made only in dark colors, which were pressure-molded into objects such as telephone and radio housings, and electrical components. In time, however, the Bakelite Corporation of America figured out how to produce Bakelite in clear, translucent, and marbled varieties in a wide range of colors. This more colorful Bakelite was perfect for making jewelry, and was usually cast into shapes such as tubes and rods, which were then cut into bangles and other pieces. Bakelite jewelry is almost never molded, and because of its thermosetting qualities, it cannot be bent, but it is excellent for cutting and carving. It has an extremely high refractive index, meaning that it can hold a brilliant shine. Phenol formaldehyde proved to be hazardous to work with, and was almost completely discontinued in the period after World War II, although a small amount is still produced for technical purposes. The contemporary jewelry designer Judith Evans uses a safer version of phenolic resin without formaldehyde (see "New Designers").

Some confusion has been caused by the existence of scores of other trade-names for phenol formaldehyde. When the Bakelite Corporation's patent lapsed, other companies jumped in with their own versions of phenol formadehyde resin. The best-known competitor is **Catalin**, which was widely used in jewelry. A common misconception is that Catalin can be distinguished by the presence of marbling. While each company's color formulas were slightly different, both Bakelite and Catalin were produced with and without marbling. More to the point, these are only the two most well-known of dozens of brand-names. Unless you have a piece of jewelry with its original tag or card, or you are lucky enough to find an original advertisement for your exact piece, it is impossible to identify the brand of plastic. We follow the common convention among collectors and use "Bakelite" to refer to all phenol formaldehyde. One exception is **Prystal**, a highly transparent phenolic resin that can be identified with some certainty. At first used for a different plastic, the name Prystal was purchased by the American Catalin Corporation in 1935. It can be found in a range of colors, including yellow (some of which was originally clear), red, blue, purple, and green. It often has a dichroic quality, revealing different colors under different kinds of light. While very popular with collectors, it is more prone to cracking and heat damage than other kinds of Bakelite.

**Urea Formaldehyde**, sometimes known by the trade name Beatl, retains its color better than Bakelite and was sometimes used for cream or white versions of Bakelite designs.

A row of floral and leaf-carved Bakelite bangles, $125-500.

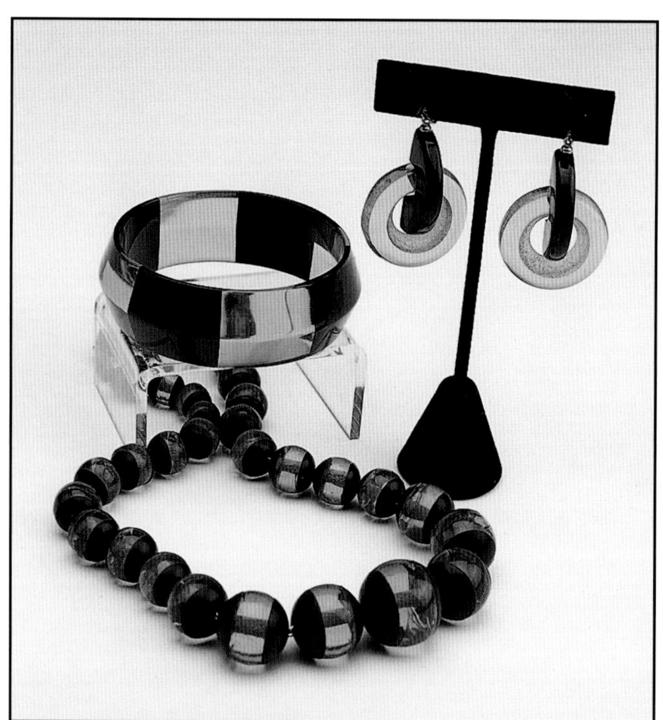

A "married" group of apple juice and black Bakelite pieces, using three different techniques for joining the colors. The bangle shows cast stripes in alternating colors, $150 and up. This piece comes from England, where this technique was used more often than in the U.S. The earrings are made of two pieces screwed together, $45-50. The beads are laminated, $95.

Stack of chunky black Bakelite bangles, $125-250.

Red and green prystal hinge bracelets with coordinating pieces. Hinge bracelets, $250-400; buckle, $45; clips, $35 each; earrings, $25.

**Lucite**, the trade-name of methacrylate resin, was patented by the Dupont Company in 1937. After the patent lapsed, it was manufactured and sold under many different trade names, but as with Bakelite, collectors use "Lucite" as the generic term. While it is a thermoplastic, it can have a very hard shiny surface. It may be either molded, or cast and carved, although for inexpensive jewelry, molding is far more common. (The same piece might be formed by molding and decorated with carving.) While Lucite bangles that have been molded may occasionally have seams, on higher quality pieces these are buffed out. The best pieces are mostly made by casting and do not have seams. Many contemporary designers work in cast Lucite and other resins. **Alexis Bittar** is well-known for his luminous hand-carved Lucite pieces (see "New Designers").

Watercolor stacks of vintage transparent Lucite bangles, $10-40 depending on size and color.

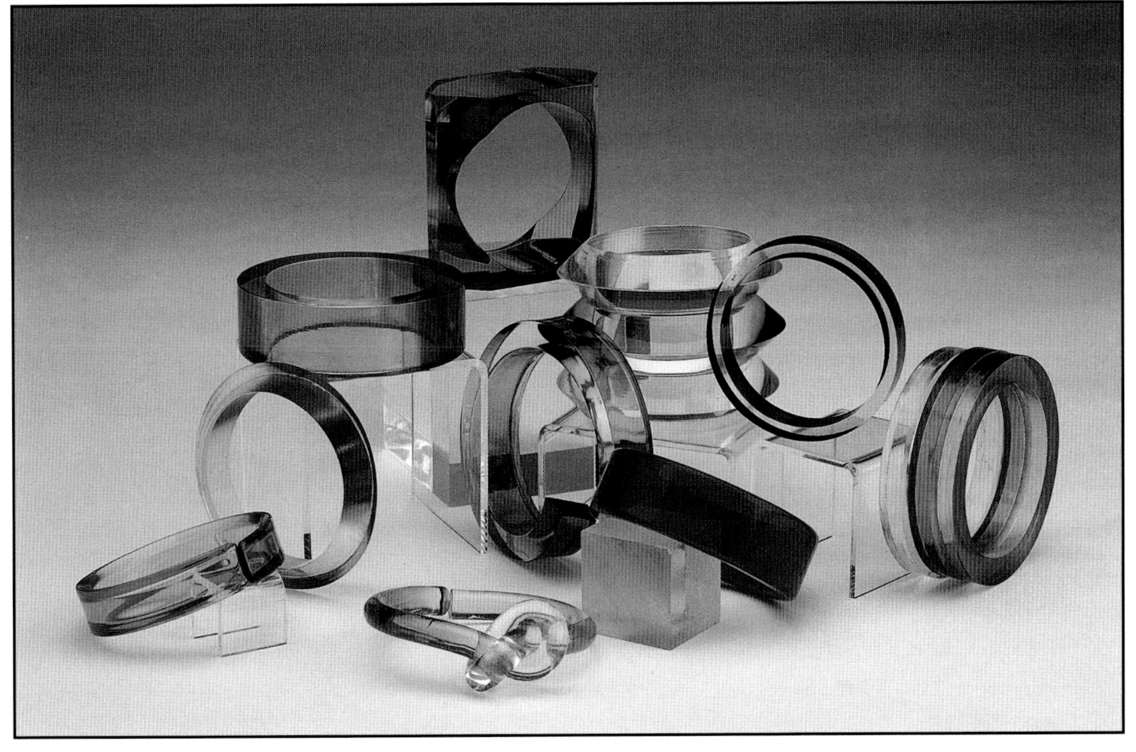

Unusual shaped and shaded Lucite bangles. The clear and green saucers are new, $10 each; the rest are vintage, $25-45.

Frosted molded and carved Lucite bangles. Clear to purple molded, $20-30. Stack of clear frosted bangles, from top: wave-carved, $55; daisy carved, $35; molded, $20; molded, $35. Wave-carved, courtesy Lori Kizer.

Pink Lucites of all types. Front stack, $20; rest, $45-60.

Vividly colored Lucite and hard plastic tube bangles, $15-20.

Stack of bangles made of an unknown material, not Lucite but heavier than hard plastic. Multicolored, $35; other colors, $10-15.

Hinge bracelet in pearlescent Lucite, showing unusual construction of beads on metal frame, with matching necklace and earrings, $30-35 for set.

**Hard plastic** is a catchall term for a number of cheap plastics that are frequently used to make bangles and other plastic items. It can often be distinguished by the presence of seams, which either run around the inside or outside of the bangle or across it at the halfway point. Another clue is the high-pitched clink when bangles rattle against one another. Older bangles will often show a dusty whiteness along their edges. Hard plastic is quite brittle and is never carved and rarely decorated by hand. It may have beads, rhinestones, or fabric glued to its surface. While a few hard plastic pieces are collectible, most are of little interest to collectors.

A set of twelve hard plastic marbled bangles in original box, "Made for Chadwick-Miller Inc., Boston, Mass. Copyright 1970. Made in Hong Kong." $10-15.

Marbled hard plastic bracelets, from top: butterscotch two-piece bracelet with dangles, $25; pink hinge, $20; red hinge, $15.

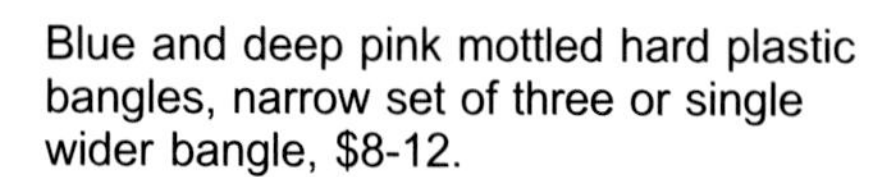

Blue and deep pink mottled hard plastic bangles, narrow set of three or single wider bangle, $8-12.

**Soft plastic** is not a technical term, but we will use it to indicate some of the plastics that were used in the production of jewelry in West Germany after the Second World War, as well as some more recent types. The Danish company Buch + Deichmann used a soft but luminous plastic during the 1980s to produce vintage-inspired pieces that are now beginning to interest collectors.

Since World War II, so many complex synthetics have been developed that – short of subjecting your prized pieces to destructive chemical analysis – exact identification is impossible.

Soft plastic bangles made in West Germany, probably 1960s, $4-8; black bangle with X-carving, unknown origin, $10.

Signed Buch + Deichmann bangle, Denmark 1981, loosely imitating vintage style, $12-15.

**Resin** is a general term embracing most plastics. Bakelite is phenolic formaldehyde resin; Lucite is methacrylate resin, etc. Some contemporary jewelry designers work with substances they identify only as resins, which are similar but not identical to Lucite and are well-suited to casting and carving. These often have a translucent, frosted appearance.

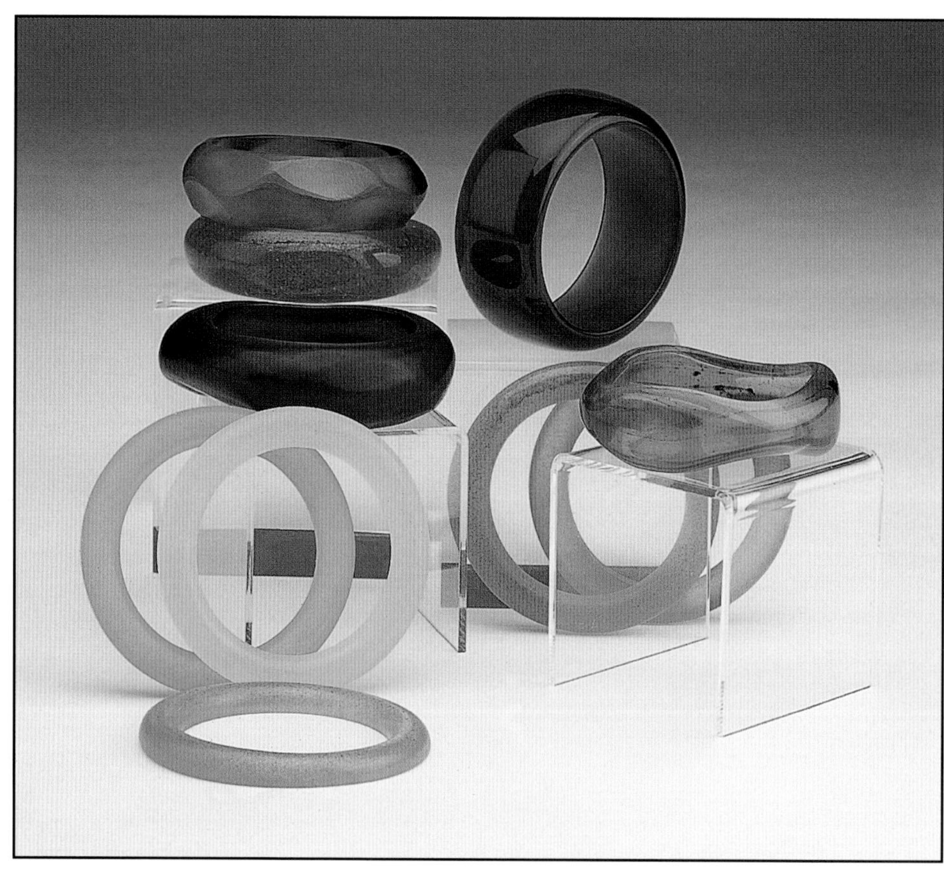

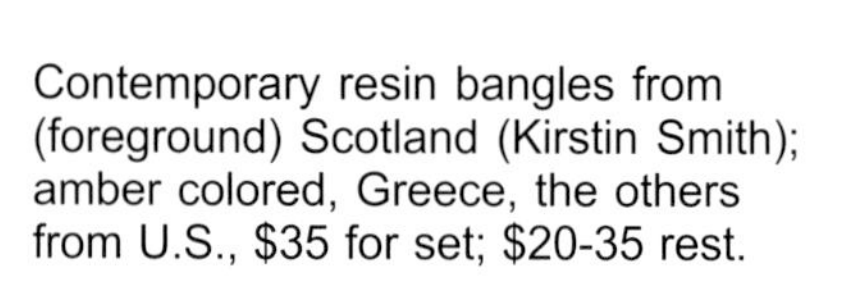

Contemporary resin bangles from (foreground) Scotland (Kirstin Smith); amber colored, Greece, the others from U.S., $35 for set; $20-35 rest.

## A Note on Periods and Styles

Dating plastic jewelry can be rather difficult. Similar styles were made over a long period of time, and newer pieces often deliberately mimic older ones. Hence we often give a range of dates. On the rare occasions when you see a single date, that is because we have clear evidence, such as a copyright on the box in which the item was originally sold.

Learning to identify the different plastics will allow you to make more educated guesses about the period of manufacture. Stylistic features can also be helpful, but remember how often older styles were imitated in new materials.

Descriptions of plastic jewelry frequently contain the terms "art deco" and "retro," which should be used with caution. "Art deco" is used appropriately of the period 1925 through the 1930s in Europe, arriving a bit later in America. It refers to an angular style that represents a departure from the flowers and tendrils of art nouveau. It is characterized by geometric patterns, and is never frilly and rarely floral. A plain bangle with no unusual shaping cannot be described as art deco. "Retro" is an especially tricky term, properly used to indicate a deliberate evocation of an earlier period. A vintage piece should never be described as "retro" unless it imitates an even earlier style.

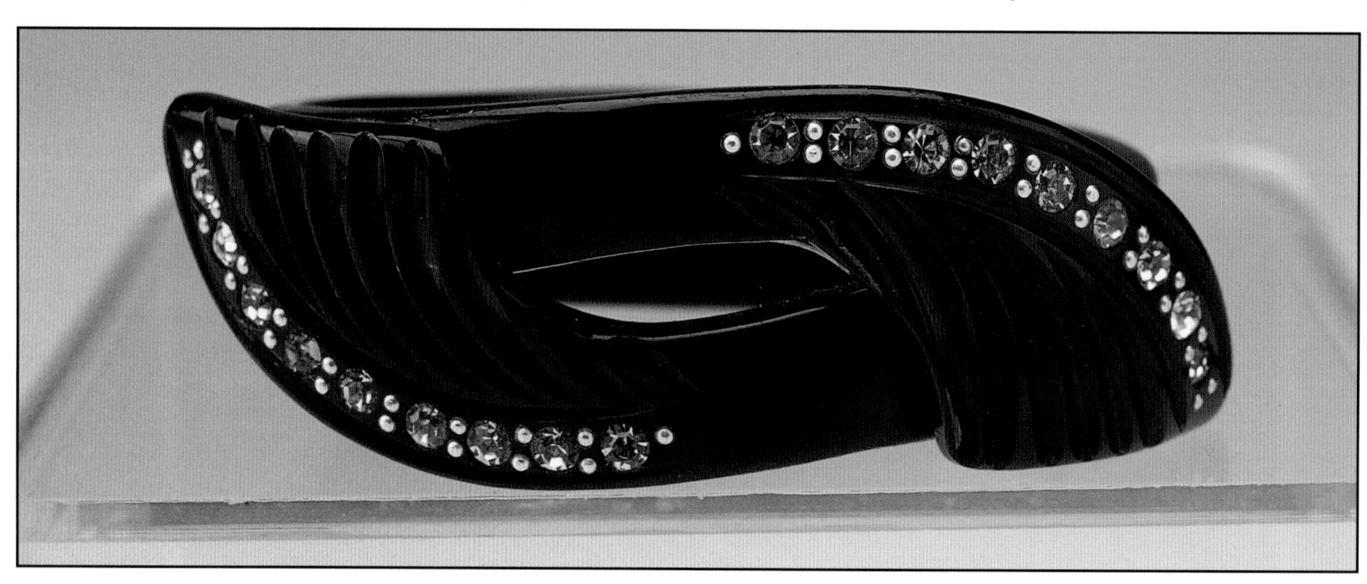

Black deco celluloid bypass bracelet with rhinestones, $85 and up.

Chapter 3

# Identification of Plastics

In this section, we aim to de-mystify one of the most vexing issues for collectors and dealers. One of the saddest phrases we hear is, "I was told it was Bakelite." Even honest and reputable dealers do not always know how to identify plastics accurately. This is why it's so important to learn the basics for yourself if you plan to buy or sell. A little knowledge can mean the difference between profit and loss, between pride and disappointment in your collection. If you remember a few basic points, you will rarely if ever have to rely on anyone else's opinion about a piece.

Below we will give you instructions on testing plastics, but our true aim is to make testing only rarely necessary. Once you have learned to recognize the properties of the main types of plastic you will no longer need to test 95% of the pieces you find. This is important because definitive tests exist only for celluloid and Bakelite, and even these are not infallible. Also, some pieces are too delicate to test or have decorative elements that could be damaged in the process.

In theory you could use all five senses in identifying plastics, although we do not recommend that you taste your bangles! Of course your eyes will do most of the work, but there is much to be learned by smelling, touching, and listening. Some of the best tests rely on the use of smell. When heated, Bakelite gives off a smell like burning insulation, while celluloid smells of vinegar or camphor, and galalith smells faintly of burnt milk. (See "Testing" below.) Lucite on the other hand has almost no smell.

Feel the weight and texture of the bangle. Celluloid and cellulose acetate are feather-weight, hard plastic is light, Lucite is heavy, and Bakelite heavier still. (Oddly enough, "fakelite" is the heaviest of all.) Unlike the other plastics, only the thinnest Bakelite is at all flexible. Celluloid and cellulose acetate are both highly flexible, and thus suited to bypass and wrapped shapes that must flex and open slightly to be put on. Lucite varies in flexibility, depending on thickness.

Let your fingers check for roughness or the presence of seams, which may indicate hard plastic. (If you are buying, your sense of touch can also prevent you from getting stuck with a cracked or chipped piece.) A single and sometimes irregular seam across the inside of the bangle may indicate celluloid, which was often stamped out in strips, bent into a circle, and joined. A very smooth inside surface, sometimes slightly concave or convex suggests Lucite or another modern plastic, while a very flat, but not shiny surface, sometimes slightly textured or with fine ridges, suggests Bakelite. Bakelite bangles may also have smooth knife-edges. Hard plastic is especially prone to slight indentations in otherwise flat surfaces.

Gently tap a few similar bangles together. If you hear a high-pitched almost metallic clink, they are hard plastic. If a bangle sounds hollow, it is probably hard plastic. Lucite and Bakelite make deeper sounds which some profess to be able to distinguish, although we have found this method inconclusive.

These hard plastic bangles all have visible seams. The wide red one on left has a seam across the bangle on the outside. The red on the right has a seam running around the inside. The black facetted one has a seam running round the outside middle. Notice also the chalky-white edges on the wavy-edged bangles. These signs of wear are found only on hard plastic.

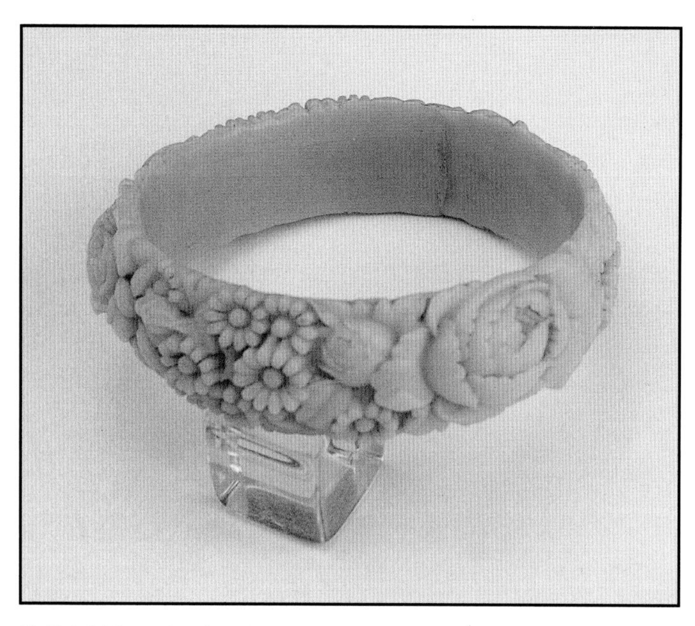

Celluloid bangle showing irregular seam across inside where the ends of the strip have been joined.

Now, examine the bangle, taking into account design, color, and method of decoration, if any. If there is a **design** in relief or intaglio, does it look hand-carved, which would indicate celluloid, Bakelite, or Lucite, or does it seem to be stamped or molded? This suggests celluloid or any of the modern plastics. Too exact repetition, especially of a highly intricate design, means that the piece was not hand-carved. If so, it is unlikely to be Bakelite, which was rarely molded. (This is one case where rarity does not make for greater desirability.) Reverse carving (sometimes with paint added) was used most often with Bakelite and Lucite. Here again, hand-work should be distinguished from machine production, as the cheaper plastics are rarely hand-decorated. Some hard plastic bangles are adorned with slapdash hand-painting, which can nonetheless be charming.

On right a pair of molded Lucite bangles (probably Monet) $5-12; left and bottom right, two molded hard plastic bangles $4-7. All of these show exact pattern repetition.

A matched pair of Lucite and Bakelite bangles, showing a convex inside wall for the Lucite (top) and a straight up-and-down wall for the Bakelite (bottom). Lucite $35, Bakelite $125.

A group of Lucite bangles designed to resemble Bakelite in both pattern and color. The two on the right appear to be hand-carved, while the rest are clearly molded, $12 and up. Carved bangles, $25-35.

A group of Bakelite bangles with imitations in Lucite, hard plastic, and even wood. From top, two thumbprint Bakelite bangles ($125; $75) with a matching cream Lucite ($45); group of three abstract leaf-patterned bangles, butterscotch, cream, and white are Lucite ($15-25), the dark red is a resin-washed Bakelite with more carved detail on the leaves ($85); shaved diamond-pattern: frosted and moonglow Lucite ($35) with dark green Bakelite in middle ($55); stack of four narrow bangles, resin-washed and dark cream Bakelite ($35-45). Second from top is stained wood ($10). Z-carved pair on right: red Bakelite ($80) and blue marbled molded hard plastic ($15). Gouge-patterned pair on left, top is cream Bakelite ($125), bottom, yellow marbled molded hard plastic ($15). Center bottom, cream Bakelite daisy-carved ($40) with matching moonglow Lucite ($35).

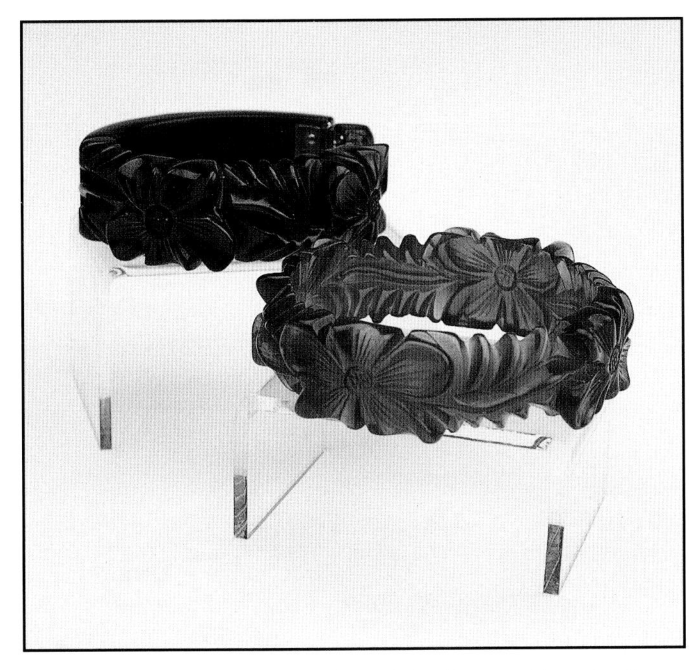

A fine carved Bakelite hinge bracelet with a molded plastic bangle in the same pattern. Bakelite hinge bracelet, $250, bangle, $25.

A group of molded bangles. Clear and blue hard plastic versions of a Bakelite rose pattern $15-20; peach translucent Lucite bangle $25-35; apple juice molded resin with seagull pattern, $35-45.

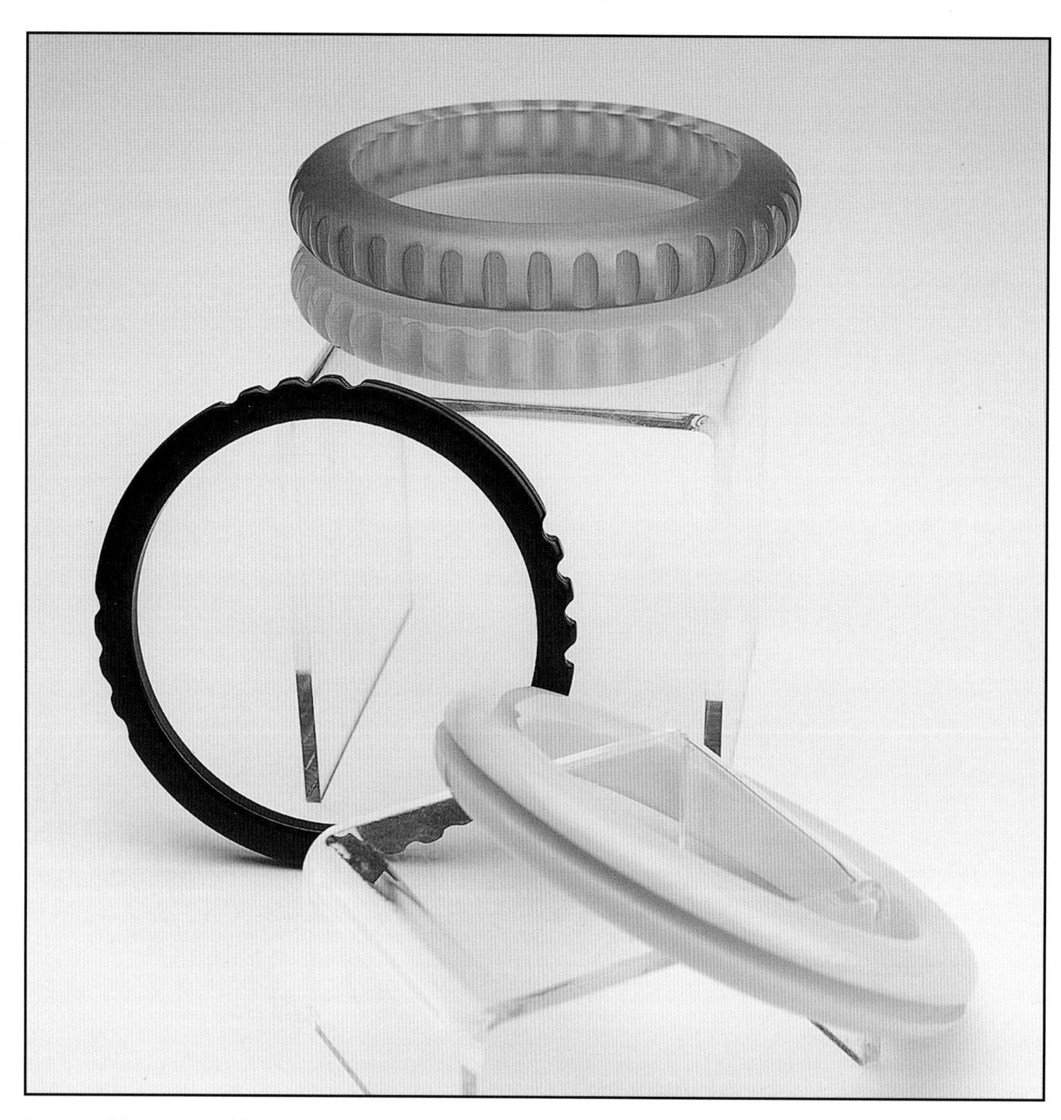

Group of four carved bangles. Although the opaque ones are sometimes taken for Bakelite, they do not test, and are made of Lucite, $10-30.

Pink and white Lucite bird-carved bangle $20-35. This pattern is usually found in Bakelite, in both vintage and newly carved versions.

Molded hinge and bangle bracelets in Lucite, imitating Bakelite. The reddish-brown one is Trifari. $30-35.

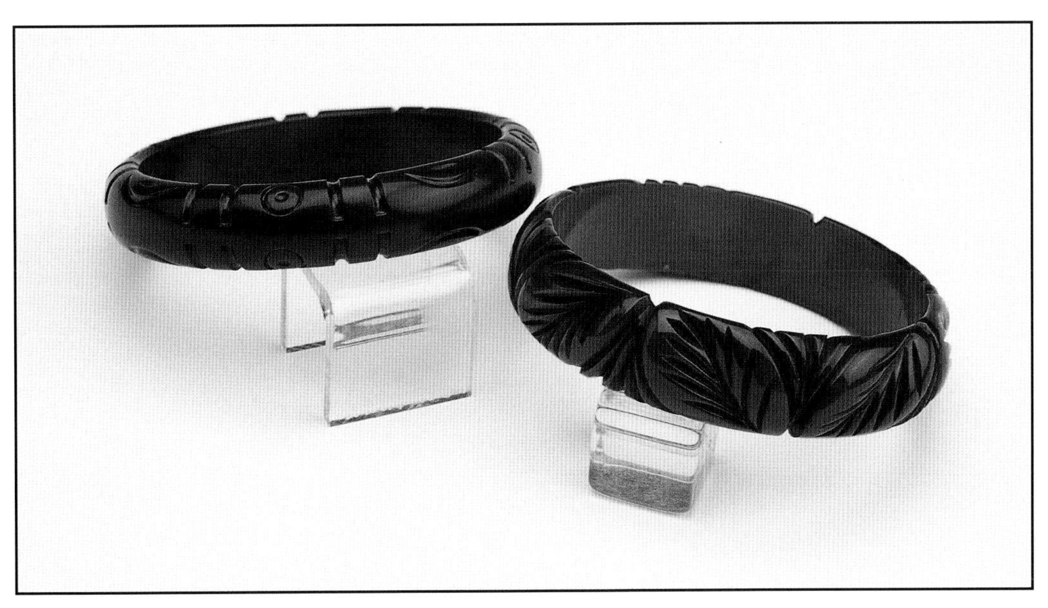

Left, molded Bakelite bangle; right, carved resin-washed Bakelite bangle. Molded $10-15, carved $150.

A pair of bangles with reverse-carved painted pink roses. The top is a Lucite hand-carved and hand-painted piece, while the hard plastic bangle on the bottom shows the regularity and shallowness of a machine-cut pattern. Top $30-35, bottom $10-15.

A group of decorated bangles from top to bottom: narrow hand-painted bone $25, medium hand-painted plastic $25-35, and wide hard plastic with decal $10-15.

A group of confetti Lucite bracelets with matching earrings and necklace. Bracelets, $25 and up, necklace, $20, earrings, $15 and up.

**Inclusions** in the plastic are extremely uncommon with celluloid and Bakelite (although Bakelite with metallic dust was made for a couple of years in the 1930s). The use of confetti glitter, shells, flowers, insects, and other fragile materials only became possible in the late 1950s with the development of cold-cast plastics. All of the plastics under discussion may be found with **applied** rhinestones, metal, plastic, paint, wood, and other decorative elements.

In the last decade or so, newly carved pieces of vintage Bakelite (as well as other plastics) have come on the market. These are distinguished by the repetition of a few patterns, somewhat clumsily rendered, and are generally found only in black, red, yellow, and blue-grey. Recently the market has been flooded by pieces imitating vintage Bakelite, although they are made of a different plastic. They appear at first glance to be hand-carved, but many of them are not. The best protection against being fooled is a thorough acquaintance with vintage Bakelite. Are the colors plausible and consistent throughout the piece? Are the designs familiar to you from pieces you have handled or from study of the many excellent Bakelite books available? (See "New Plastics" for an expanded discussion.) Remember that a lot of inexpensive pieces were made to imitate pricier materials, so that similarity of pattern alone is not enough for identification. At the other end of the price spectrum, a few skilled modern artists are producing new designs in old Bakelite or similar materials. Most of these pieces are decorated with lamination rather than carving, they are signed by the artists, and are quite different in feeling from vintage pieces.

Is there a **label**, **stamp**, or designer's **signature**? Each of these can provide clues. Many of the celluloid bangles sold in the U.S. were stamped on the inside in purple ink, "Made in Japan." While this tends to wear off, traces can often be found. Many Lucite and hard plastic pieces still have a small paper label saying where they were made, usually W[estern] Germany, Hong Kong, Taiwan, India, or China. (This is excellent evidence but not conclusive, as a label may end up on a bangle that it doesn't belong to.) Some European pieces have a signature molded in relief, such as those by Buch + Deichmann of Denmark. Bakelite bangles almost never carry a signature or statement of origin, but in any case were almost entirely produced in North America, Britain, and Europe. **We know of absolutely no Bakelite jewelry produced in Japan or India.**

Higher quality Lucite pieces often carry a small plaque with the name of the company. A hand-written incised signature tells you that the piece was individually made, and is most likely Lucite or another modern resin. An inscription by the giver or owner of a piece is usually amateurish enough to be easily distinguished from an artist's signature. The vast majority of bangles are unsigned, but some of these can be identified by a matching pin, necklace, or earrings bearing a trade name stamped into the metal findings.

A floral celluloid bangle showing a "Made in Japan" stamp.

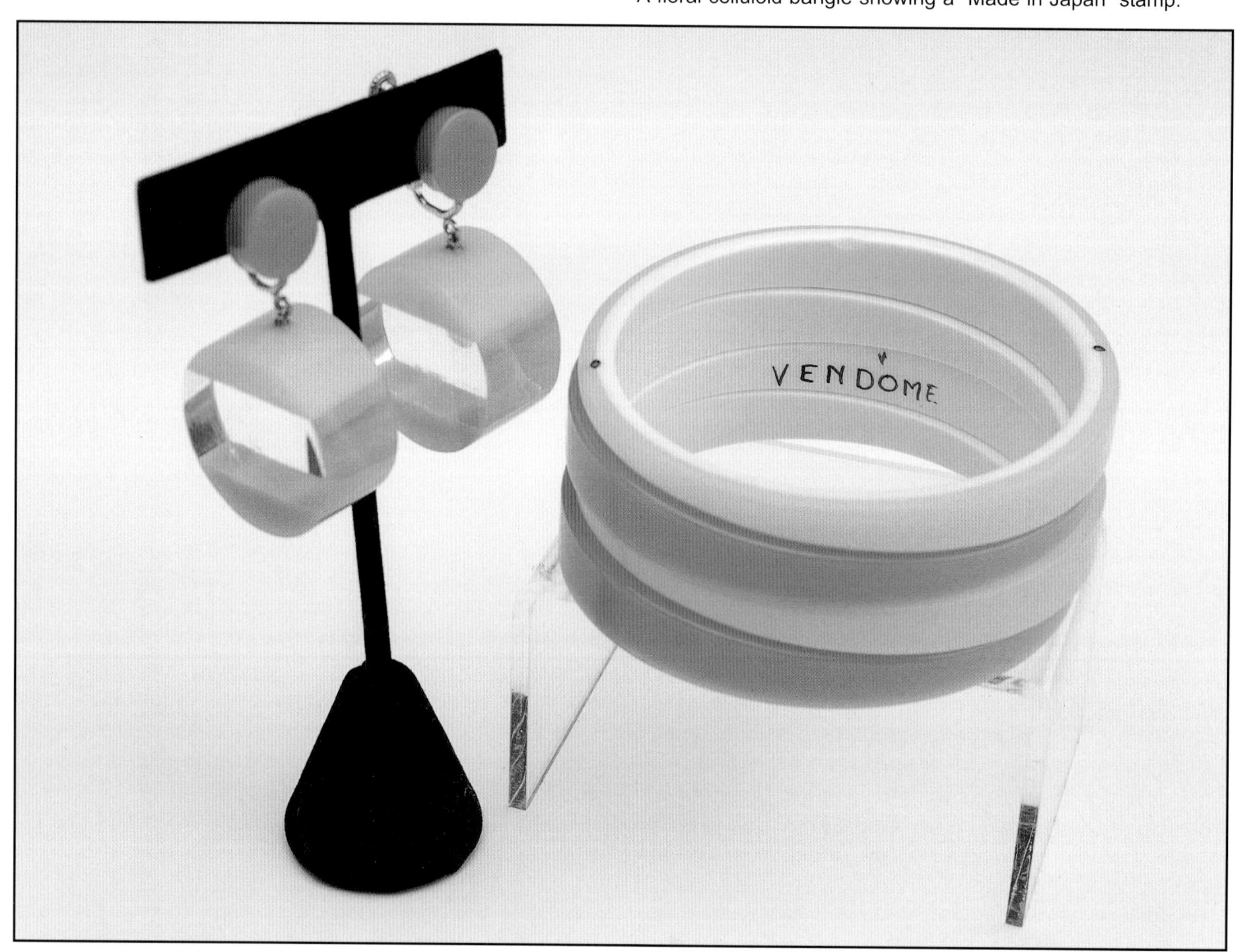

Earrings and a bangle, both signed Vendome. The earrings are signed on the metal earring backs. The bangle is unusually constructed of four separate pieces held together with a metal pin on each side. Earrings, $15-20, bangle, $25-35.

A beautiful reverse-carved Lucite bangle signed by the artist, "John v[an] d[er] Heide Rio," $95.

Sparkle celluloid bangle inscribed "H. L.," presumably the owner's intials.

True apple juice Bakelite, heavily carved. The carving creates a frosted look, but the uncarved parts are a clear transparent yellow-gold. Bangle, $250-350; matching earrings, $35; clip, $25-40.

**Color** is important in distinguishing among plastics. When describing colors, you may want to use the terms set by Karima Parry in her book, "Bakelite Bangles." If you aren't sure what to call a color, describe it as you see it. Remember that opaque colors allow no light through, translucent colors allow some light through, and transparent colors are completely "see-through." If you can see your finger moving behind the bangle when it's held to the light, it's translucent. If you can read the newspaper through it, it's transparent. The color "apple juice" has caused some confusion lately, so if you aren't sure, take a look next time you are in the supermarket. (If it's cloudy like unfiltered apple juice, it can be called "applesauce.") Apple juice Lucite or Bakelite is absolutely transparent and light to dark yellow in color. Other transparent colors are sometimes called "lime juice," "cranberry juice," etc. Transparent orange is sometimes called "tangerine juice" rather than "orange juice" to avoid confusion. Prystal – as far as we know – was only made in completely clear, not marbled, colors.

Celluloid tends to come in cream, ivory, and other pastel colors as well as black. Celluloid floral bangles can be either coated with a single dark color or tinted with several colors to highlight leaves and flowers. However bright the tinting may once have been, most of the colors are now pale. We have heard of "artists" now painting celluloid bangles in bright colors, which seems a shame, since the original delicacy of the coloring is lost. Given this, we recommend caution when buying celluloid florals with multiple deep colors, especially as they command high prices. Celluloid is sometimes found in a pearlized pattern that can be confused with Lucite. Pearlized celluloid – usually a thin sheet laminated to a clear or translucent layer – shows up in vanity items (mirrors, brushes, etc.) more often than in jewelry.

Celluloid floral bangles, molded, showing a variety of patterns, including lily-of-the-valley, and elephant, two in unusual shades of green, $35-45.

Group of floral celluloids showing degrees of tinting. The colors on the top bangle and the large chrysanthemum cuff have mostly worn off. Bangles, top to bottom $25-45; chrysanthemum bracelet, $50 and up.

Plastic hinge bracelets sold under the name "Featherweight," mid-1950s. These imitate celluloid, but have a precision of detail that is not found on celluloid floral bracelets. Also found in pastel colors, and with rhinestones added. For plain white, $15-20.

Imitations of celluloid floral bangles in hard plastic and Lucite. The garish colors and aurora borealis coating give away the hard plastic ones. Celluloid floral bangles are occasionally found in translucent colors, but rarely if at all in transparent ones. Large transparent Lucite, $15-20; others $7-10.

Openwork bracelets. Top, unusual pearly and clear layered celluloid overlap bracelet, perhaps meant to be worn on the upper arm, $125; center, newly made cellulose acetate cuff, $18; bottom, cream celluloid open bangle $125, courtesy Nancy Schiffer.

Celluloid bangles, from top: carved, $40; bamboo patterned bangle, $25; plain, $10; wide imitation tortoise-shell bangle (real tortoise-shell would have a seam), $25; thin bangle with beads, sized for the upper arm, $35; thin bangle, $10; black cut back to cream, $35-45.

Cellulose acetate comes in a wide variety of colors. The designer Lea Stein uses layers of cellulose acetate in different textures, colors, and patterns that give her pieces a distinctive appearance of depth and luminosity. Lea Stein's pins and other pieces are usually signed, but her bracelets are not. Other producers have used less complicated effects, but may also work with layered sheets. Some cellulose acetate is still produced but it is relatively uncommon.

Group of cellulose acetate bangles and cuffs with matching pieces, old stock from a factory.

Bakelite was originally made in a wide range of colors, but over time the lighter shades have darkened considerably. Familiarize yourself with the characteristic colors of Bakelite and you will save yourself a lot of guesswork (and money). White Bakelite is almost unheard of, as it has all darkened to cream or yellow. Only refinished pieces or those kept in a light-free environment will be white. Blue, pink, and purple exist but are hard to find, as they have mostly darkened to murky greens and browns, so pieces in these colors identified as Bakelite should be scrutinized carefully. You may occasionally come across a piece that has been refinished, giving you a chance to see if you prefer the original colors or the deeper, sometimes richer ones found today. Over the span of a year or two, these will revert to the darker oxidized colors.

Three floral and leaf-carved Bakelite bangles in typical colors. Note piercing in red bangle. $100-175.

A group of Prystal and other transparent phenolic bangles. Narrow, $20-40; wide, $45-60; carved, $125 and up.

Two rare teal "changer" bracelets, a hinge and a heavily deeply carved bangle. Changers show a different color when held to the light, revealing the brilliance under the oxidated outer surface. These two become brilliant teal blue. Hinge, $150; bangle, $300-350.

Translucent and transparent marbled Bakelite pieces. The lighter blue-green and purple ones are most likely French. The rest are American. Bangles, $35-80; earrings, $25-40.

Brilliant teal Lucite carved bangle $35-45; applejuice carved Bakelite bangle $300 and up.

All of these bangles were sold at low prices as Lucite, but they are all Bakelite. Clockwise from top: carved tapered bangle $100 and up; facetted bangle $175 and up; applejuice and black striped $150 and up; saucer with rhinestones $40-45; reverse carved painted $175 and up.

As noted above, Bakelite colors – whether opaque, translucent, or transparent – may be solid or marbled. Marbling is achieved by the addition of white (now turned yellow) or black to a solid color. Other combinations of two colors have been given catchy names like bluemoon, creamed spinach, or tequila sunrise (see Parry, *Bakelite Bangles*). These should not be confused with **End of Day**, which is properly used of combinations of three or more colors. These mixtures take their name from the fact that workers combined whatever colors they had left at the end of the day, although some combinations may have been created deliberately.

Two true End-of-Day Bakelite bangles in varying degrees of translucence, $80; $150. On left, courtesy Leonore Moog.

Bakelite was also occasionally sold with a pearlized coating, which is usually partly worn off. As far as we have been able to determine, however, pearlization was not incorporated into the Bakelite material itself. This means that any piece of truly pearlized plastic is not Bakelite, but celluloid, cellulose acetate, or Lucite. Hard plastic that appears pearlized is usually coated.

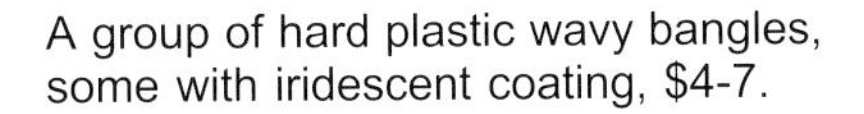

A group of hard plastic wavy bangles, some with iridescent coating, $4-7.

Galalith is found in a range of colors, some similar to those of celluloid and some resembling Bakelite. Later galalith pieces often have shaded colors applied in a wash or airbrushed onto the plastic.

Galalith bangle with shaded color-wash, French. $100 and up.

It is a common misconception that Lucite is always clear or transparent. In fact, like Bakelite, Lucite comes in nearly every imaginable color, ranging from water-clear to opaque. Unlike Bakelite, however, Lucite retains its colors with little change. (Clear Lucite sometimes takes on a slight pinkish or yellowish tinge over time.) Occasionally a laminated piece is mistakenly identified as containing both Bakelite and Lucite because one color is transparent and the other is opaque. In practice, these two plastics were not usually laminated together, and it is more likely that the entire piece is of the same substance. (Part Bakelite and part Lucite pieces do exist, but the pieces are usually glued or bolted together rather than laminated.)

Watercolor stacks of Lucite bangles.

Lucites in deep colors and a variety of shapes.

Bangle, hinge, and cuff in delicate transparent shades. The cuff is laminated. $30-45.

Stacks of colorful opaque Lucite bangles $5-15.

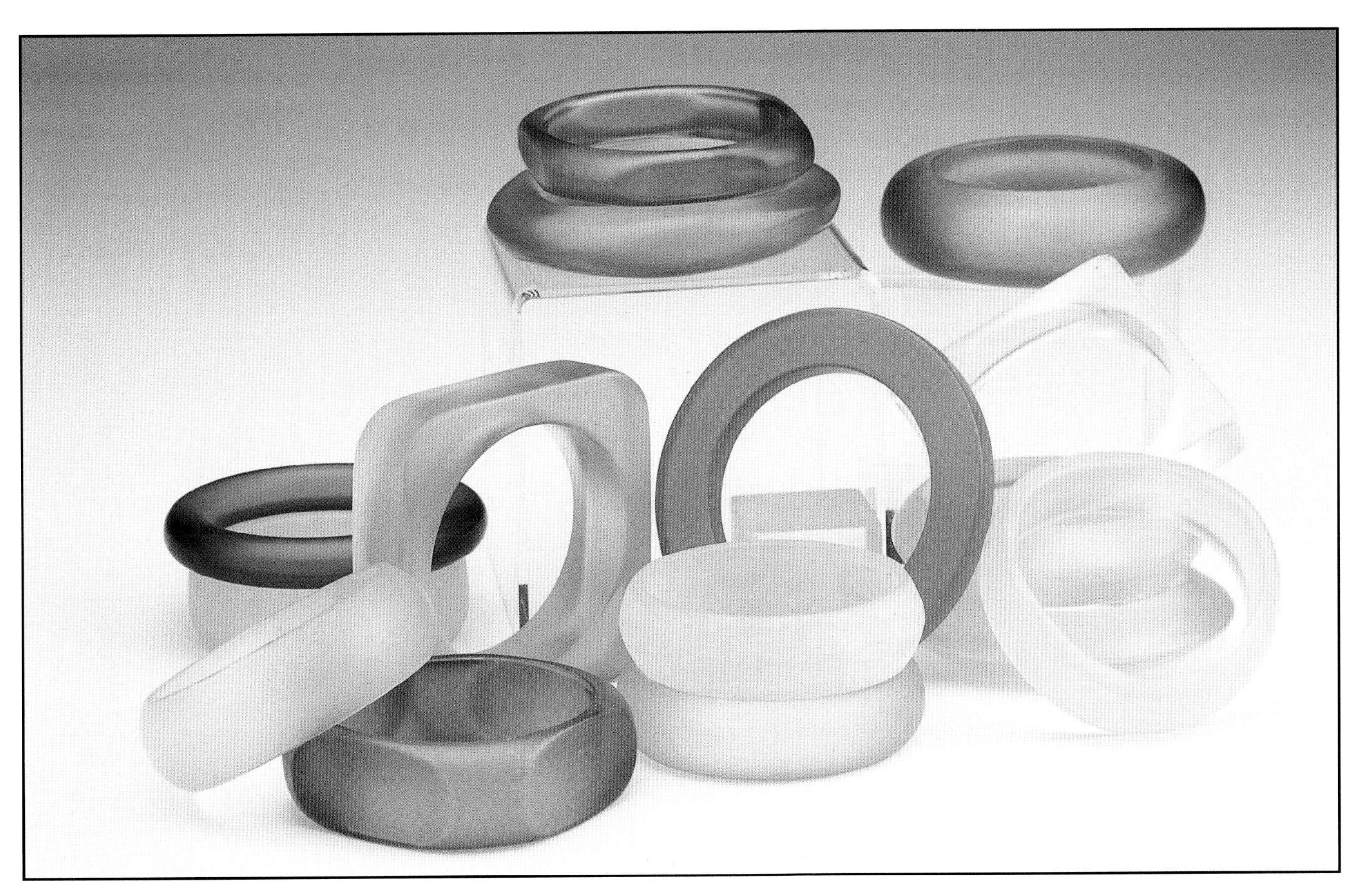

Frosted Lucite bangles, $10-45.

Not only is Lucite found in a riot of colors, loud and soft, but it takes pearlization beautifully. This effect is created by mixing powdered iridescent shell into the plastic. Pearly Lucite can be found in bangles, beads, and a wide range of other jewelry. It was very popular in the 1950s and 1960s, and is now enjoying a revival. Collectors generally refer to this effect as "Moonglow" after one of the many names under which it was sold. Moonglow has a smooth appearance with a "cat's eye" (chatoyant) effect, but a more turbulent mother-of-pearl pattern is also found.

Chunky and satiny moonglow bangles. The large gold one has a label: "P.C. Designs." $40; $15; $8.

Lucite often has marbling, which can lead to confusion with Bakelite. Here the stability of Lucite colors is helpful: if the marbling is white, the piece is almost certainly not Bakelite (unless it has been recently refinished). When attempting to distinguish Bakelite from Lucite, pay special attention to the color of any marbling, and to the saturation of the colors. The pattern of marbling can also be quite informative, as Bakelite marbling is completely random, while Lucite marbling usually occurs in set patterns.

Lucite has been used for everything from high-end hand-carved or painted designer pieces to cheap stamped-out mass produced items, so the range of possible decorating techniques is nearly endless.

Three yellow marbled bangles. Note the difference in the marbling. From top, Bakelite $45, Lucite $25, hard plastic $15.

Marbled Lucite and hard plastic bangles, some resembling Bakelite, $10-25.

Tortoise-shell bangles showing characteristic marbling patterns of different materials. Clockwise from top right, Lucite domed bangle, with marbling lines radiating out in a regular pattern, $20; sliced Bakelite bangle showing irregular marbling, $25; wide hard plastic bangle, with lines around the bangle, clearly divided into halves, $15; bottom of stack, hard plastic bangle with marbling lines around the bangle, $10-15; top, Lucite Givenchy bangle, $35.

Marbled Lucite bangles in Bakelite colors; the tubular one is hollow and probably not Lucite, $10-35.

Honey and green marbled Lucite bangles, easily mistaken for Bakelite, $15-35.

**Hard plastic**. The colors of hard plastic, especially when marbled, tend to be muted compared to both Bakelite and Lucite, but solid colors are similar to those of Lucite. The surfaces are often duller, as these plastics do not take or keep as high a polish as Bakelite or Lucite. Because of the cheapness and fragility of these plastics, labor-intensive handwork is rarely lavished upon them.

**Soft plastics**. These come in a range of colors, from pale to brilliant and highly saturated. The plastic used by Buch + Deichmann in the 1980s has a slight translucence, while the soft plastic pieces produced in West Germany in the years after World War II are usually opaque, and somewhat flat in color.

Left, stack of Bakelite spacers, made in France. Right, hard plastic marbled spacers. Note the saturated color of the Bakelite in comparison to the hard plastic. Bakelite $15-20, hard plastic $3-5.

Three similarly constructed bangles using plastic elements on a metal frame. From top, Bakelite $20, middle, hard plastic, $10-15, bottom, Lucite $15-20.

Hard plastic marbled bangles in green, yellow, and orange, $10-20.

Pink and blue marbled hard plastic bangles. Note the wide "bluemoon" in the middle, not to be confused with bluemoon Bakelite, $10-20.

## Testing Methods

**First of all, never use a hot pin!!!** At worst you could be injured, as celluloid can explode when heated. At best you will get an inconclusive result, and you may end up damaging a good vintage piece in the process. Even Bakelite can be marked by this method, although the pin will not penetrate.

The hot pin test is in any case unnecessary, as Bakelite is the one material that can almost always be identified by other means. We use the following tests: Simichrome, rubbing, and hot water. Simichrome is a high-quality metal polish made in Germany, which is excellent both for polishing Bakelite and for testing. It is pink, but turns mustard-yellow when rubbed on Bakelite. Most pink metal polishes will react the same way, but Simichrome is among the less abrasive ones. Be sure to test on the inside of the bangle, using a very small amount of polish, and wash it off afterwards. (Dow's 409 cleanser can also be used, but it won't polish your bangle.)

The polish reacts with the oxidized surface of the Bakelite, so objects that have lost their oxidation through refinishing will not test positive until the oxidation builds back up over time. European Bakelite gives off a slightly darker color, while fakelite gives off a color that is darker still. Be careful not to confuse the color of the piece coming off on the rag with a positive test result, especially with brown and yellow pieces. It is often said that some colors, notably black and red, do not test, but this is an overstatement. A small percentage of these do not test, but most will. Even if they fail the Simichrome test, red and black Bakelite should still pass the sniff test. Conversely, wood bangles may give a false positive for Bakelite, if phenolic resin was used as glue in the laminating process.

The "sniff" or "smell" test is carried out either by rubbing the object very hard until your finger feels warm and then smelling the finger, or by holding or dipping the object in very hot water for a minute and then smelling the object. (Smokers should use the hot water method.) In either case, you will get a pungent odor like burning insulation. It is important first to learn what Bakelite smells like, as other plastics also give off strong and unpleasant odors. Celluloid smells like vinegar or camphor, while galalith gives off a faint smell of burning milk. Some claim that Lucite gives off a petroleum smell, but in our experience, it usually has no smell at all. The same is true of hard plastic. The new fakelite pieces give off a strong and unpleasant odor, but not the right one. If they are left for awhile in water and then retested, the odor will become less and less like that of Bakelite. The water may also become discolored.

The rubbing test is the most convenient and can be carried out discreetly while you are shopping. It is also unlikely to cause any damage to the bangle. Use both the Simichrome and hot water tests selectively. If a piece has been dyed, Simichrome will take some of the dye off, which can make it less attractive, reduce its value, and also give a confusing result. Be especially careful with valuable transparent pieces that may have a colored stain. Also avoid using water on any piece with rhinestones, leather, wood, or cloth decoration. If you must, use a small amount of Simichrome away from the decoration and wipe immediately. When you are in doubt about whether testing will damage a piece, don't do it. Review the identification criteria above and rely on your senses of sight, touch, and hearing instead.

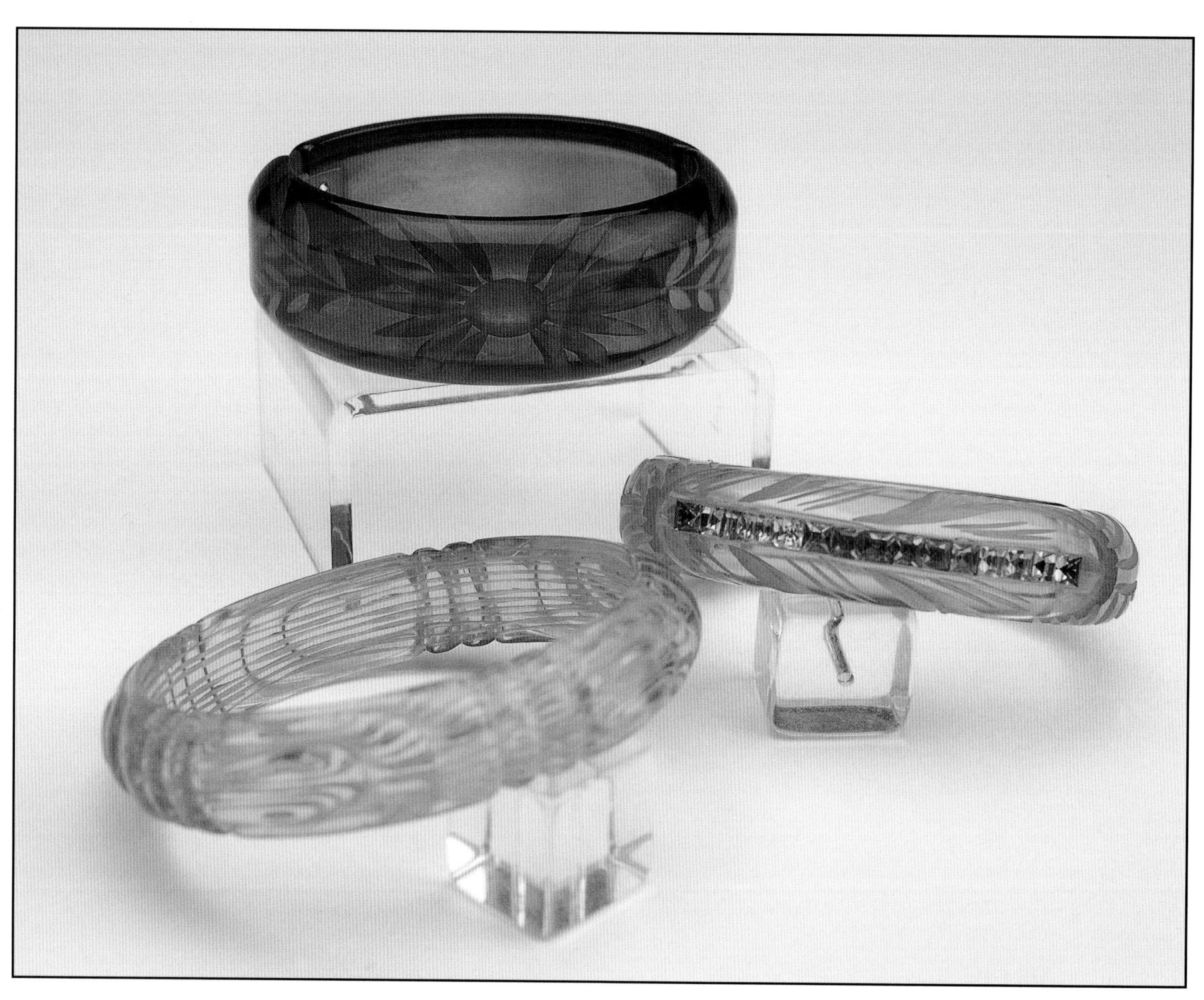

Group of transparent Bakelite (or Prystal) bracelets. The pink-lavender stained one is an example of the kind of bangle that should not be tested using Simichrome or hot water, as the color will be damaged. Rubbing a fingernail inside the carving produced the familiar phenolic smell. Bracelets with rhinestones should be tested carefully on the inside and wiped clean. Pink hinge, $175 and up; small apple juice hinge carved with rhinestones, $150 and up; apple juice carved and reverse-carved, $100 and up.

## Other Materials

Bangles are made of many materials besides artificial plastics. Keep in mind the possibility that what you have is made of wood, shell, papier-mâché, clay, ceramic, enameled metal, ivory, horn, reconstituted amber, agate, jade, glass, etc. Painted wood bangles can often be identified by a characteristic tan color on the inside, as well as their light weight. Keep in mind that bangles decorated with mother-of-pearl have a smooth shiny surface that is sometimes mistaken for plastic.

Temperature is a great aid in distinguishing plastics from other materials, as plastics (including amber) are usually warm to the touch, while glass, bone, stone and other hard organic materials are often cold. Don't forget your common sense — if it doesn't look or feel like plastic, it probably isn't.

Group of horn bracelets (snakes and openwork) $15-25; with Lucite and other plastics designed to imitate horn, carved $25-35; marbled (French), $35; cream and dark brown (French), $20 and $40.

Group of wooden bangles with bone inset elongated dots, some dyed black, $20-45.

Group of bracelets made from natural materials including wood, bone, and Hawaiian kukui nuts, $10-25.

Group of bangles in assorted materials, including stone, glass, cinnabar, mother-of-pearl, and wood inset with bone. Large wood with bone $25 each, others $10-15.

A joyous array of painted or stained wooden bangles, collectible in their own right. Painted wooden bangles can often be spotted by a distinctive tan color on the inside. $20 and up.

Stack of black and white painted wood bangles $35-45.

Russian papier-mâché handpainted and signed, "Nina," $35-50. Courtesy Joan Young.

Set of red amber necklace an
stretchy bracelet; reconstitute
amber bangle, $10 (center);
Egyptian glass bangles, $2-4.
All contemporary except the re
amber, which is most likely
vintage, $65 and up.

Stone bangles and imitators, clockwise from top: black marble, Lucite, glass (pale green), jade, and agate, $15-20.

A colorful stack, from top: mother-of-pearl on wood, mother-of-pearl on wood or plastic core; dyed bone on wood $20-35.

(Clockwise) Ceramic bangles, enameled metal, plastic bangles wrapped in plastic or cloth tape, set of three pink, purple, and black anodized aluminum, and two metal bangles with a layer of enamel on the outside, $10-25.

Cinnebar bangles with two hard plastic imitations interspersed. The wider, darker ones are real, $15-20.

An ivory bangle (white, center), surrounded by imitations in a variety of materials. The openwork elephant bangle is made of bone, the one below it is celluloid, and the rest are hard plastic. This group shows how celluloid was intended to mimic ivory, and was in turn imitated by other materials. Ivory, $100 and up; celluloid $35; bone $20; others $10-20.

Chapter 4

# TYPES OF BRACELETS

**Bangle** – a rigid closed circle (occasionally square or oval) pulled on over the hand, found in almost all materials. Not a synonym for bracelet, it is properly used only of rigid bracelets that encircle all or most of the wrist.

Group of three bangle bracelets, red carved Lucite with black paint, hard plastic openwork bangle, and laminated black and white cutback bangle with "hard carved" tag, $15-25.

Black hard plastic bangle with an interesting facetted pattern, $10-15.

**Open bangle** – a bangle that has a small opening in the circle, most often found in Lucite, occasionally in celluloid. This style allows for slight expansion of the bracelet for ease in putting on, and is found only in flexible plastics. Laminated fabric bangles are frequently made this way to avoid interrupting the design with a visible seam in the cloth.

A group showing the difference between open bangles and cuffs. The two in the front are cuffs. $15-35.

**Cuff** – a "C"-shaped bracelet, found in nearly all materials, whether flexible or not, but most common in Lucite and cellulose acetate. The term is sometimes used wrongly of other types such as bangles and hinge bracelets, but the usage recommended here is recognized by nearly all collectors.

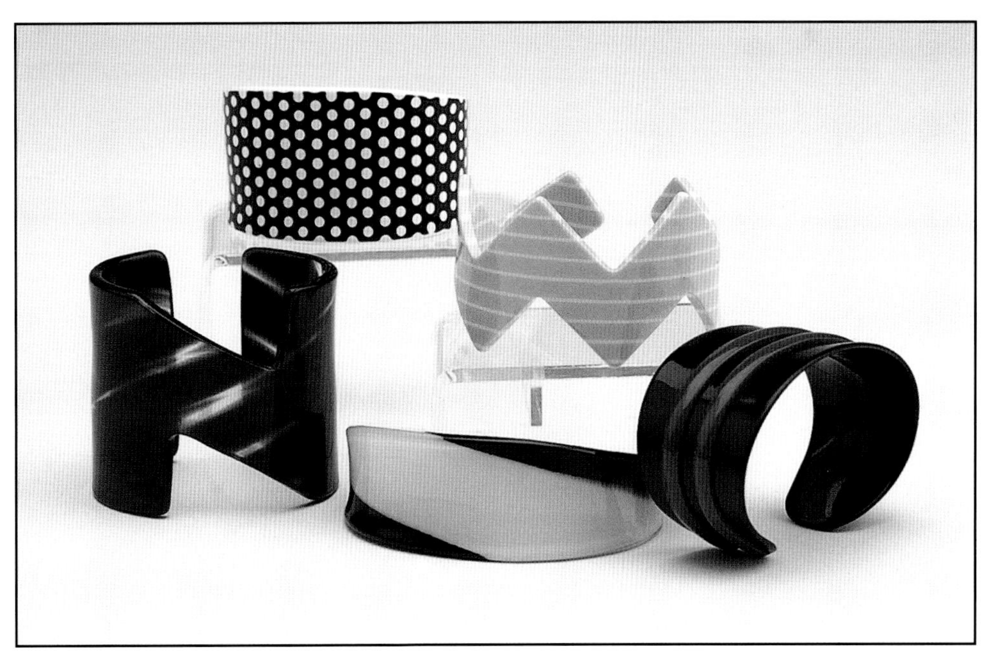

Group of cuffs in interesting patterns and shapes, including zigzag, $10-25.

**Bypass** bangles are open bangles in which the ends are angled so as to be side-by-side. These may be flexible or rigid.

A group of bypass bangles with cuffs. From top, layered bypass bangles, $20-25; cuffs, red and white, $18; orange and yellow, $8; pink and white bypass, $20-25; front row, bypass, $12-15.

Group of tortoise-shell cuffs, one with cutouts and matching earrings, $25-45.

Other variants on the bangle or open bangle include spirals and other wrapped shapes, which can only be made of non-thermosetting plastics, like Lucite. Sometimes one end is drawn through a hole in the other end of the bangle or the ends are overlapping, which requires a degree of flexibility in the plastic. This type is found most often in celluloid and cellulose acetate.

A group of matching overlap or wrapped bangles in saturated colors, $10-15.

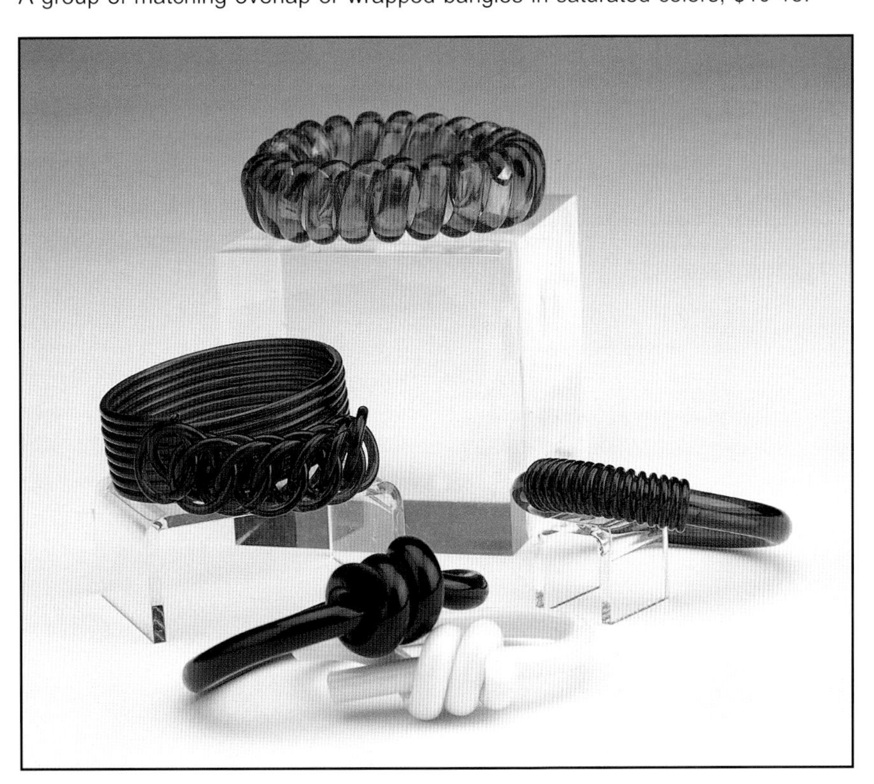

Group of bangles incorporating twisted elements. Clockwise from top, "telephone cord" bangle, $15-20; French wrapped Lucite, $15-25; pair black and white wrapped Lucite bangles, $15-20 each; French tortoise bangle of wound plastic tubing with curlicue front, $25-30.

**Stretchy** or **Elastic bracelet –** a bracelet made up of small elements, such as beads or plaques with holes running through them, strung on elastic. Stretchies come in all materials, but are most commonly executed in cheaper ones. The exception to this is the elegant and valuable transparent reverse-carved and -painted Bakelite or prystal bracelet. Stretchies can be worn by almost anyone, but (true to their name) tend to get stretched out and need periodic restringing.

Red, white, and black hard plastic and Lucite stretchies, $10-20.

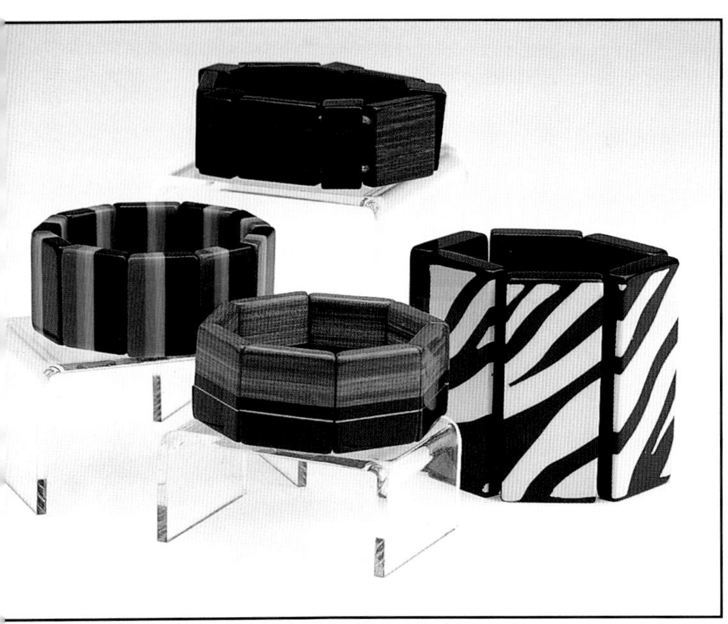

Laminated stretchies, some incorporating wood, $15-25.

Lucite and hard plastic stretchies, $10-15.

Three Bakelite stretchies and one celluloid. Bakelite, $75-150; celluloid, $35-45.

**Hinge bracelet**, also known as a **clamper** – a bracelet made of two rigid curved parts, connected by a hinge. The opening may be on the top center of the bracelet or at the side. Hinge bracelets have been popular in many periods, and include carved Bakelite show-pieces, simple Lucite and hard plastic pieces, and celluloid and Lucite adorned with rhinestones like those made by Leru and Weiss. Most hinge bracelets have a somewhat oval shape. Rarely were vintage hinge bracelets completely round. Some recent imposters have been made by slicing a bangle into two halves, smoothing one set of ends, and placing a hinge at the straight ends. This produces an inauthentic totally round hinge bracelet. (Very occasionally, vintage bangles are found that have been doctored in this way to accommodate a wearer with large hands.)

Hinge bracelets in the shape of interlaced fingers, in opaque and moonglow Lucite, $25-35; set with matching earrings, $45.

Group of hinge bracelets, most of hard plastic. Two bow hard plastic, $10-15; purple, black, and white Lucite, $10-12; Orange and white West German, $15-20; wood-look, $10-12.

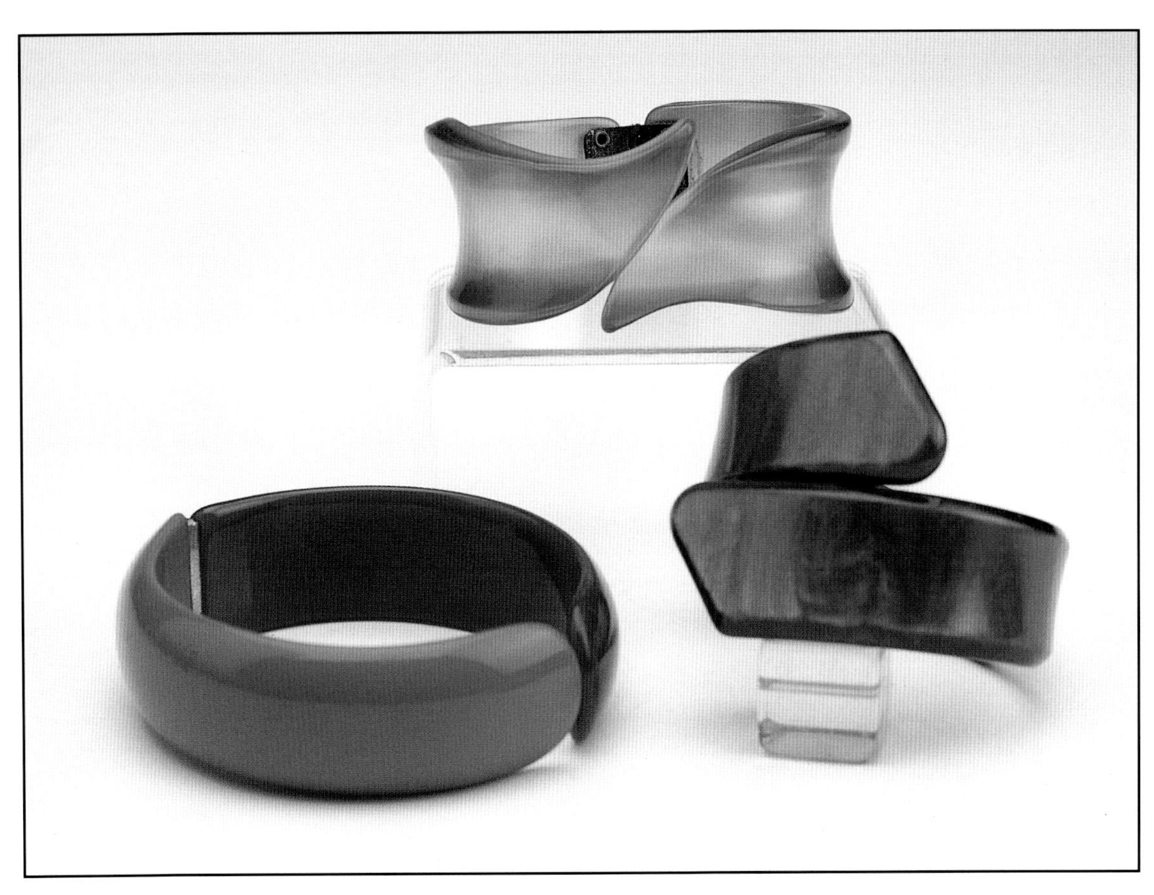

Three green and blue-green Lucite hinge bracelets, $15-25.

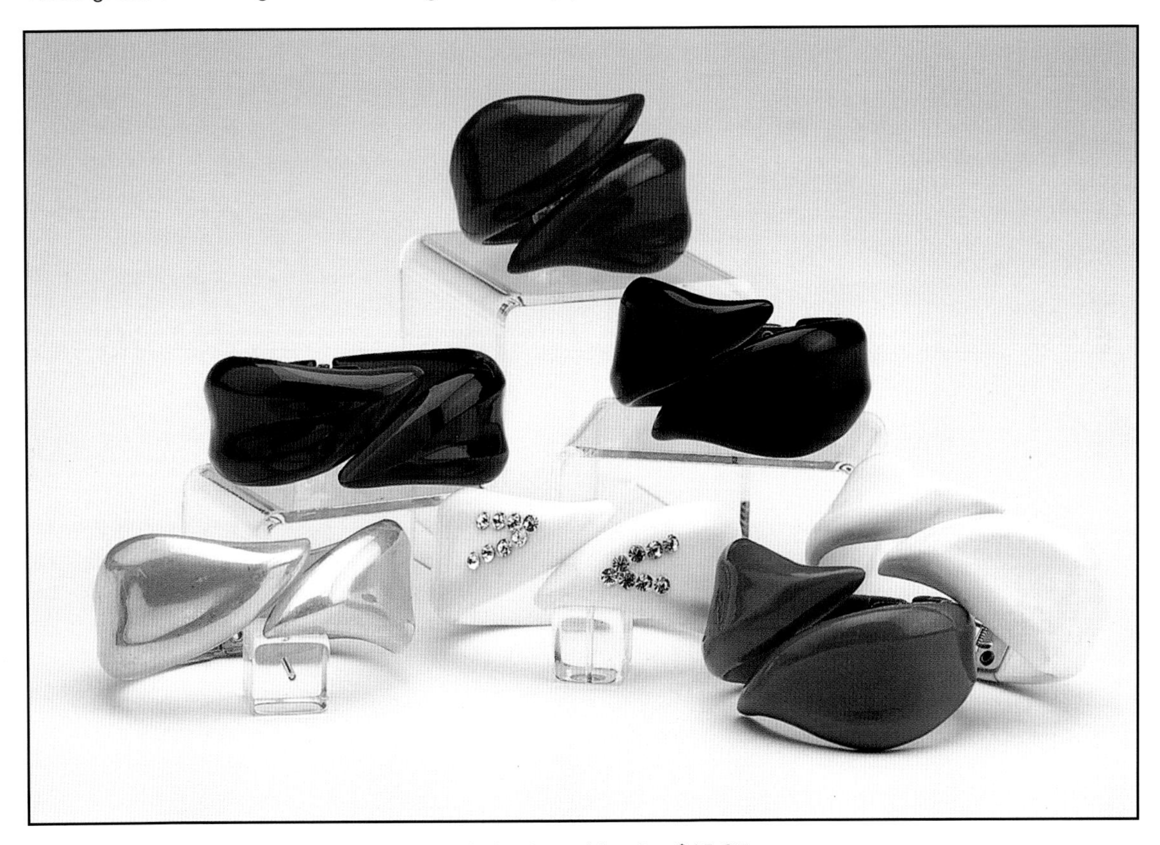

Group of leaf-shaped hinge bracelets, hard plastic and Lucite, $15-25.

**Link bracelet** – a bracelet made up of plaques or other elements linked together and closed with a clasp. These often have more metal on them than other varieties and are found in all plastics.

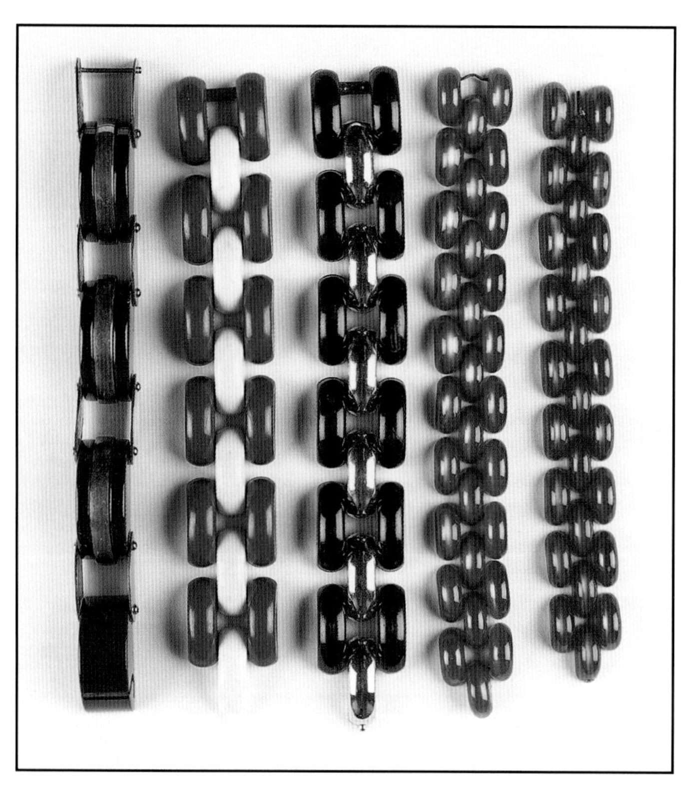

Link bracelets of Bakelite, hard plastic, and moonglow Lucite, $65; 10; 35.

**Memory Wire** – beads are strung on a wire that retains its circular shape. Some memory wire bracelets are made up of two or three rows of beads and have an opening like a cuff or open bangle, while others are made up of one long beaded wire that wraps several times around the wrist. This type often has a free-hanging bead at each end, which tends to get detached from the bracelet over time.

Three varieties of memory-wire moonglow bracelet. The neon orange and white ones wrap around the wrist, while the purple fits like a cuff, $10-30.

**Expansion bracelet** – not to be confused with the stretchy, an expansion bracelet has a metal band that expands like a watch band, to which are attached beads or other ornaments. Most often used with Lucite and hard plastics, rarely if at all with Bakelite.

Hard plastic or Lucite on metal expansion bracelets, $10-15, earrings, $8.

Light blue iridescent hard plastic "berries" on an expansion bracelet, $30-35.

**Size** – When measuring a bracelet, keep in mind that length has no meaning except in the case of link and charm bracelets, which have clasps and can be opened and laid flat. All other types should be measured by width (as they lie on the arm) and thickness (measured from the inside opening to the outside). The standard inner width of the bangle opening is 2 1/2 inches. Smaller bangles are sometimes called "maiden" bracelets. Smallest of all are baby bangles. Many of the older celluloid bangles were made in the 2 1/4 size. Avon sold many of its bracelets in both standard and large sizes. Some contemporary designers make slightly larger bangles (2 5/8 is common) but an early twentieth-century piece larger than the standard size may have been intended as an upper-arm band.

Although bangles are occasionally given a numbered size like rings, this system is not widely used and will only mystify most people. Any deviation from the standard 2 1/2 inch opening is an essential part of an accurate description. For cuffs, there is no standard inner measurement, but large and small pieces should be noted, as here the fit is more critical. Too small and it won't go on; too large and it won't stay on.

Three sizes of celluloid bangle: upper-arm band, standard, and baby bangle, $40, $30, $30.

Child-sized layered cuff, possibly Lea Stein, $25; very small Lucite cuff with rhinestones, $25-30 with all stones intact; very unusual striped bangle, $15-20.

*Below:*
A range of sizes of baby and maiden bangles in Lucite and hard plastic, some with rhinestones, $4-12.

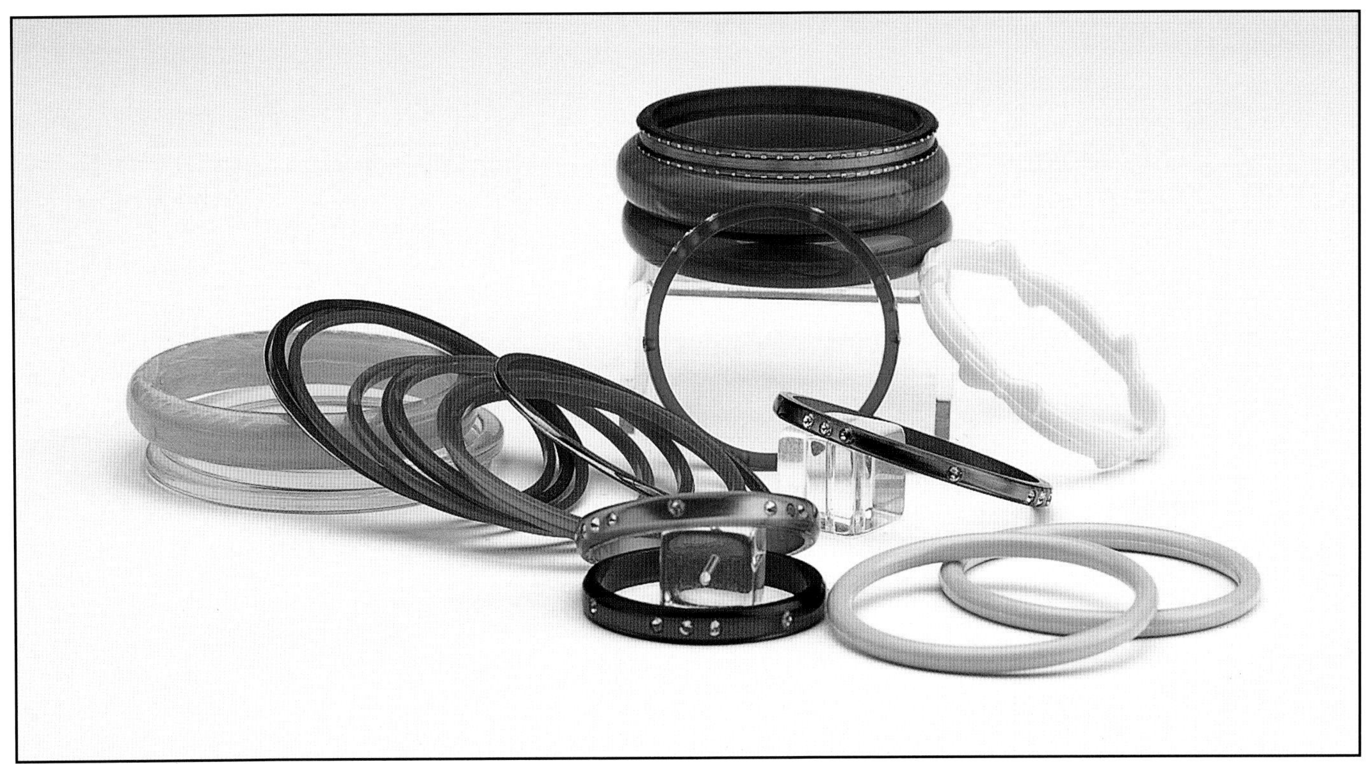

Chapter 5

# The Bangles

A rainbow collage of square transparent Lucite bangles, $5-8 each

## Shapes

The standard bangle is a perfect circle, all in one plane. (Lying on a flat surface it will touch that surface all the way around.) The inside surface is usually flat, and the outside most often **domed**. The dome is more or less pronounced depending on the ratio between the width and thickness of the bangle. The top of the dome may be flattened or beveled. Bangles which are not domed but straight up and down as if cut from a tube are referred to as **sliced**. A sliced bangle may have relatively sharp edges, or they may be rounded or beveled. A bangle whose cross-section is a circle or oval is called a **tube** bangle. A bangle whose outer profile comes to a point is called a **saucer** (as in "flying saucer"). Pairs of bangles are sometimes found that are domed on one side and flat on the other. These are meant to be used as end-pieces for a stack of sliced bangles.

Bangles may also be oval, triangular, square, rectangular, or free-form. Polygonal forms such as pentagons, hexagons, and octagons are fairly common and can be very attractive. A variant on the polygon is made by "rotating" the form, so that the points are not lined up the same way on each side of the bangle. This allows for a kind of triangular or wave-like decoration on the outer surface. A square bangle becomes an octagon by cutting off the corners. This shape can be modified by rounding the corners, and by beveling the sides in various ways. A slanted bangle resembles a cross-section of a tube cut on the bias. Other variations include (usually thin) bangles with scalloped or pointed outer edges. Occasionally you will find a bangle whose outer edges take on the shape of a face, a flower, or other object.

Once freed from the single plane, bangles can have edges that are wavy, taper, or come to points. Freeform bangles are often irregular in width. So-called "Puzzle" bracelets have complicated shapes that fit together when worn in a stack. Bangles may be flat on one side and cut on the diagonal on the other side. These usually come in pairs that fit together.

Freeform multicolored bangles lying flat.

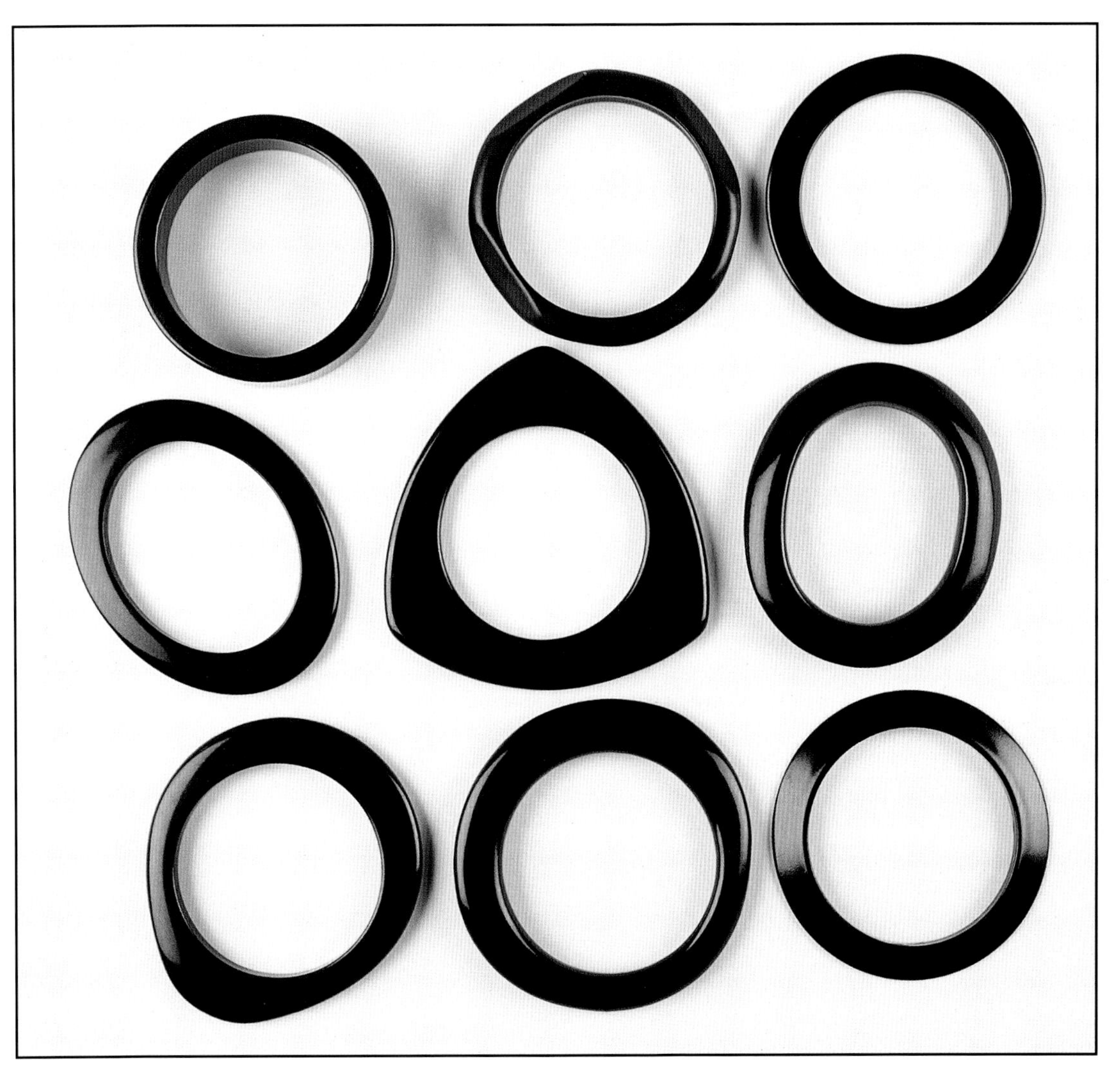

Group of black shapes lying flat.

Stack of 6 tapered freeform bangles, $8-10.

Assorted geometric bangles.

Heavy geometric bangles shot from above.

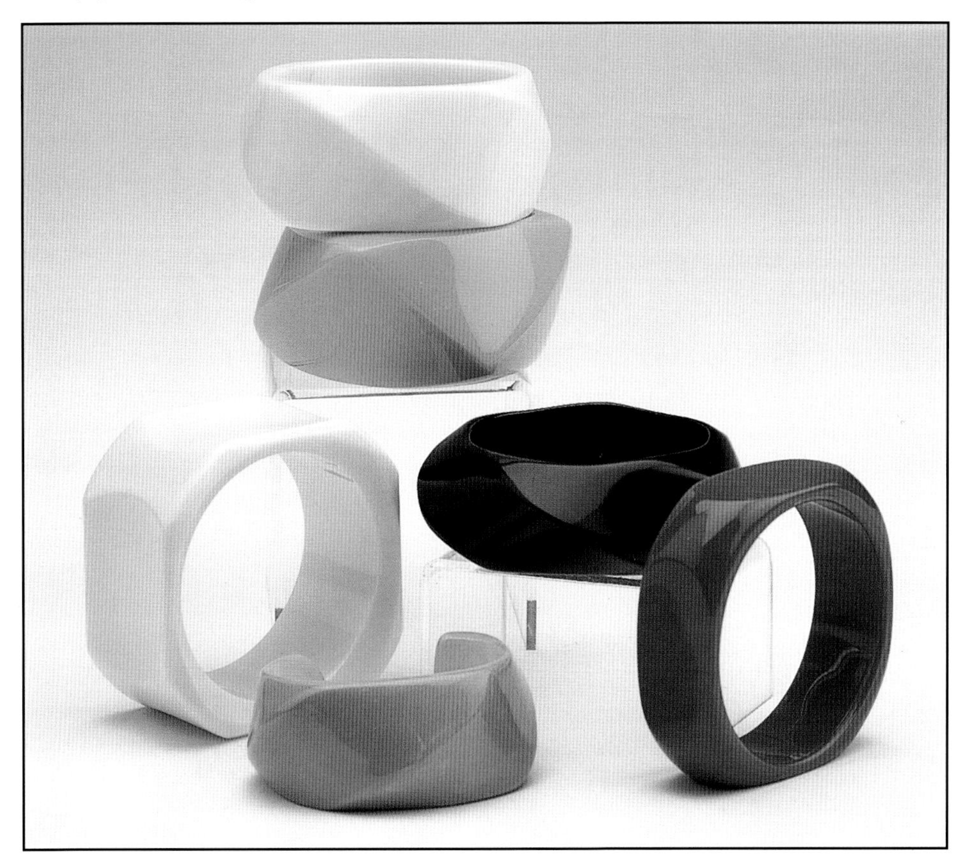

Same bangles stacked, $25-45.

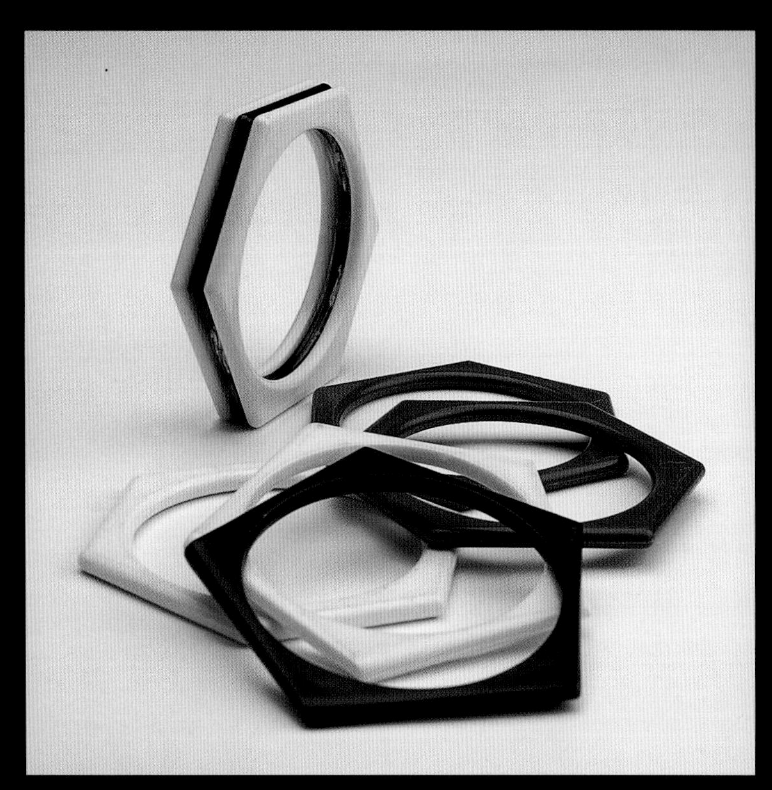

Pentagons and hexagons in hard plastic, $4-8.

Black bangles in various shapes including an unusual hand-made star bangle. $20-35; star, $45-55.

Stack of tortoise-shell Lucite bangles with an X-shaped bangle, $10-15.

Transparent pale lavendar, bent circle Lucite bangle, $10-15.

Celluloid bangles in spiral and snake shapes with rhinestones, $35-75.

Fanciful shapes: top, children's bangles; bottom Avon Curves bangles, $8-12.

Group of fancifully shaped lucite bangles, including star-shaped, saucers, and undulating, $10-15.

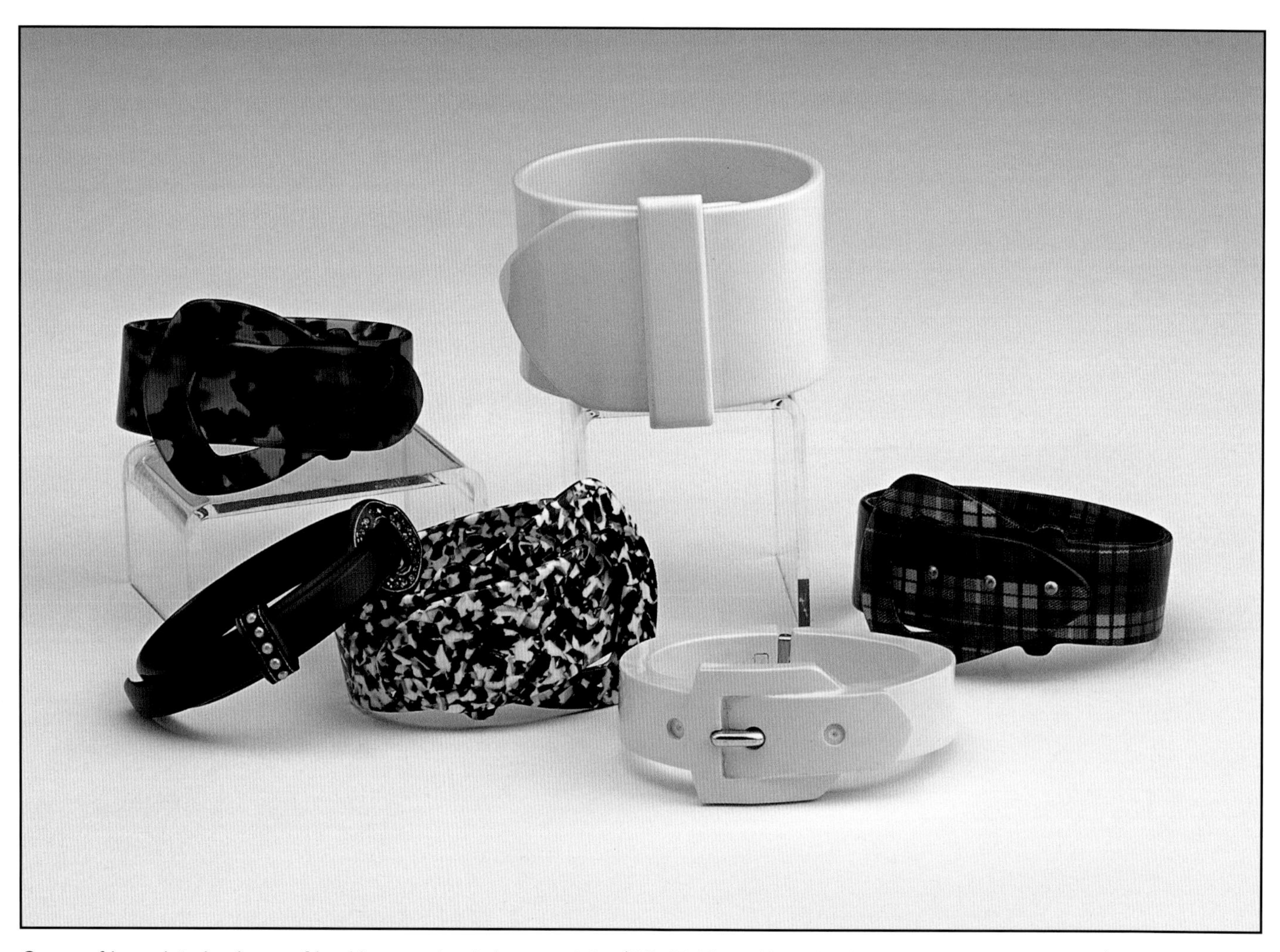

Group of bracelets in shape of buckles, most cellulose acetate, $25-45. The white with the metal buckle is Lucite, $20-25.

# Design Elements and Decorative Techniques

### Stripes, Dots, Diagonals and other Laminates

Stripes and dots in white, black, and primary colors. The yellow and dark blue are from Finland, 1960s. Bangles, $15-35; earrings, $8-20.

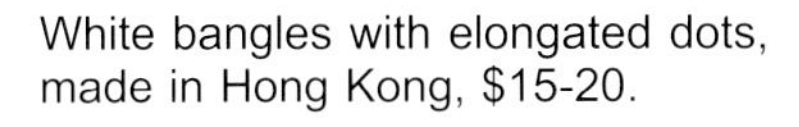

White bangles with elongated dots, made in Hong Kong, $15-20.

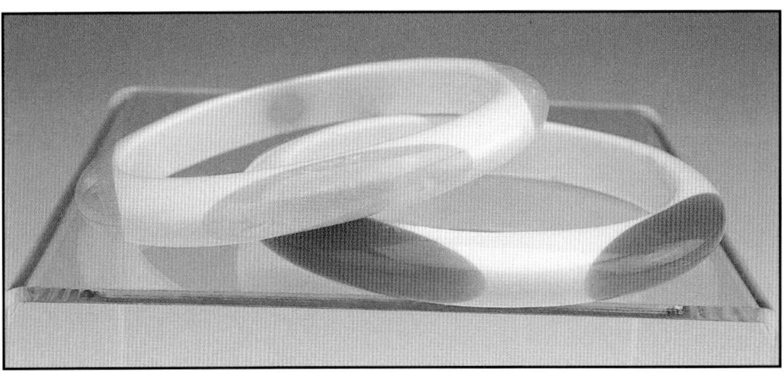

Similar bangles, but with translucent elongated dots in unusual colors, $20.

Array of vintage and new dotted bangles using different techniques including inlay, onlay, inclusion, carved and painted, etc.

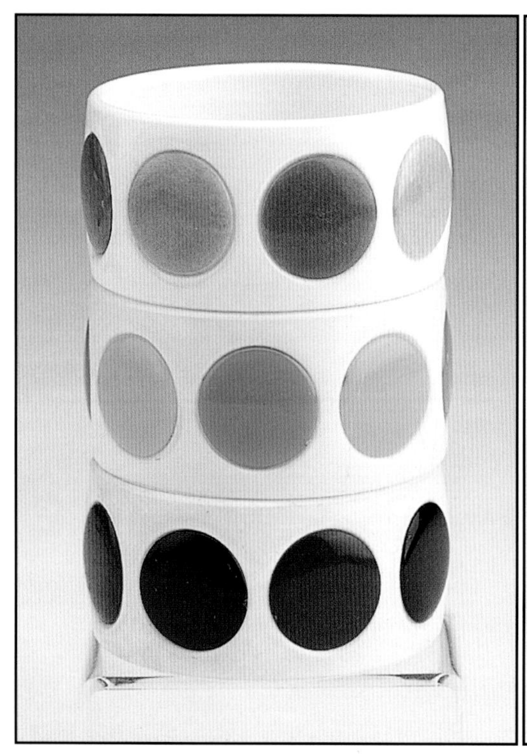

Hard plastic bangles with applied dots, $15-20.

Black and white, and red and white striped Lucite bangles, $25-35 each.

A variety of striped cuffs, $15-40.

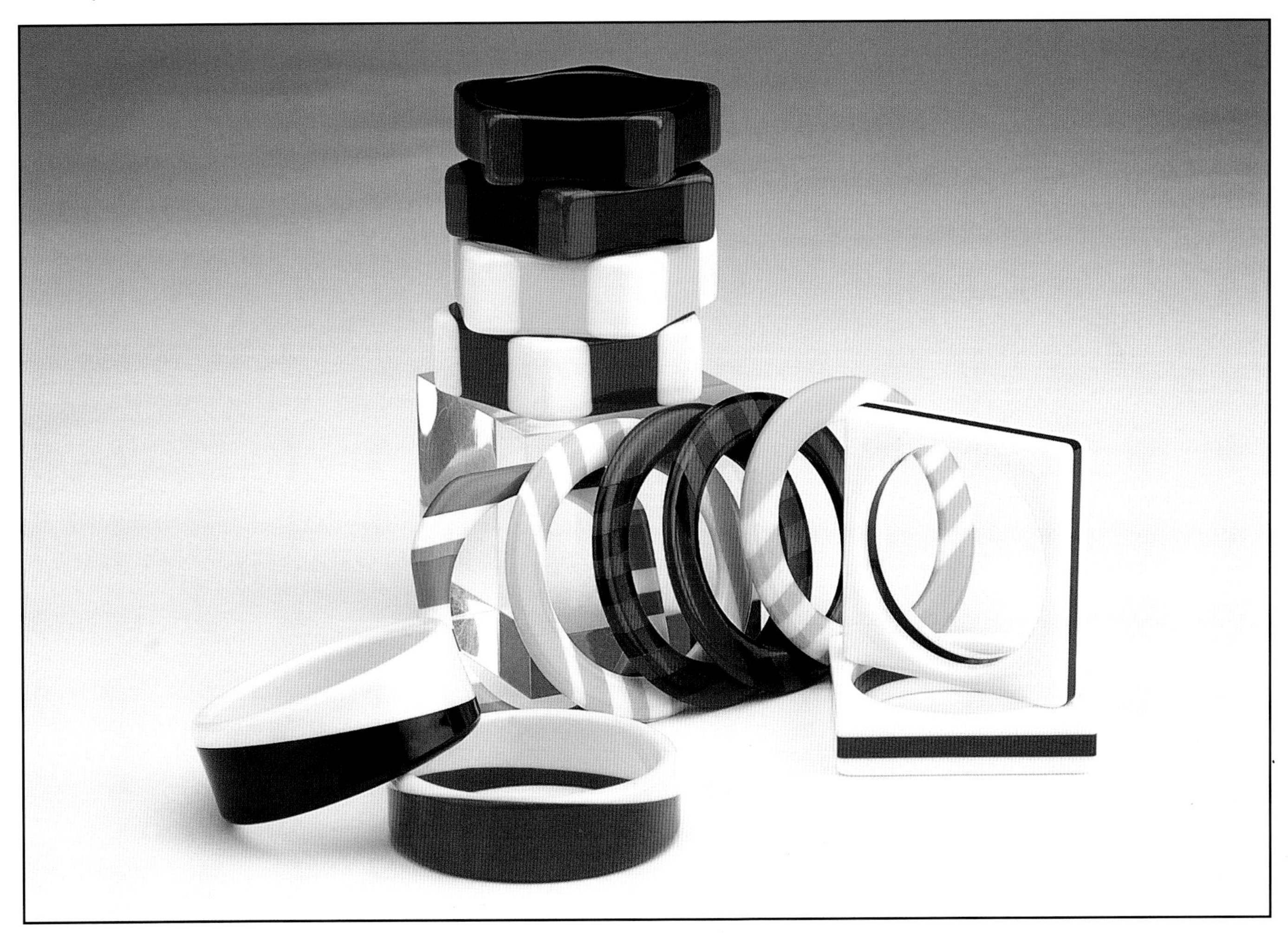

Bold laminates in white or black with strong geometric feel, $20-45.

Heavy and chunky Lucite laminates in a range of bright colors, $35-45.

Group of French Lucite bangles. Laminate; $65; blue and clear; $35; pair of hexagons, $35.

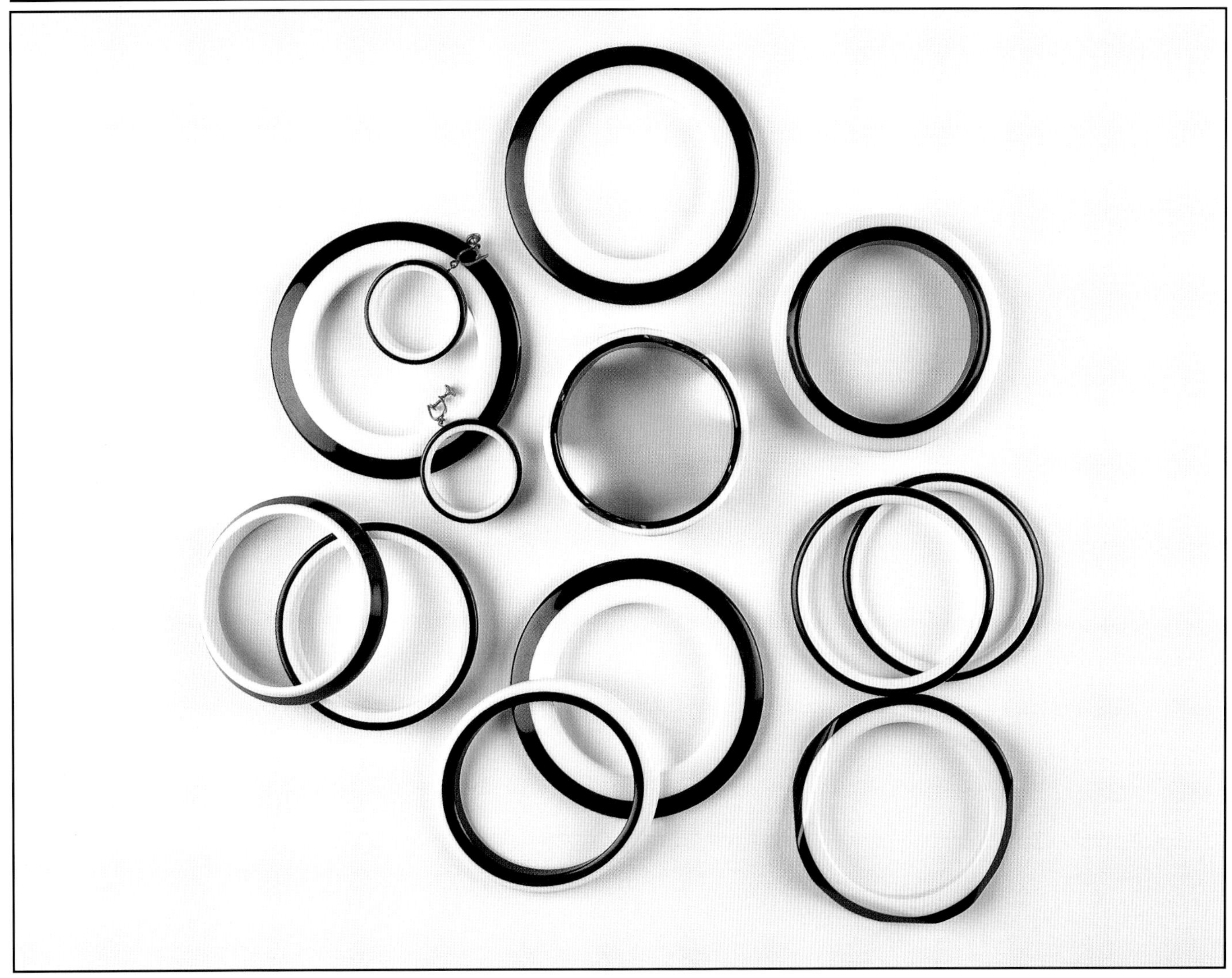

Black and white laminated bangles with earrings, $10-25.

Laminated cutback Lucite bangles and matching earrings, $15-45.

Red, white, and blue laminated cutback Lucite bangle, $35.

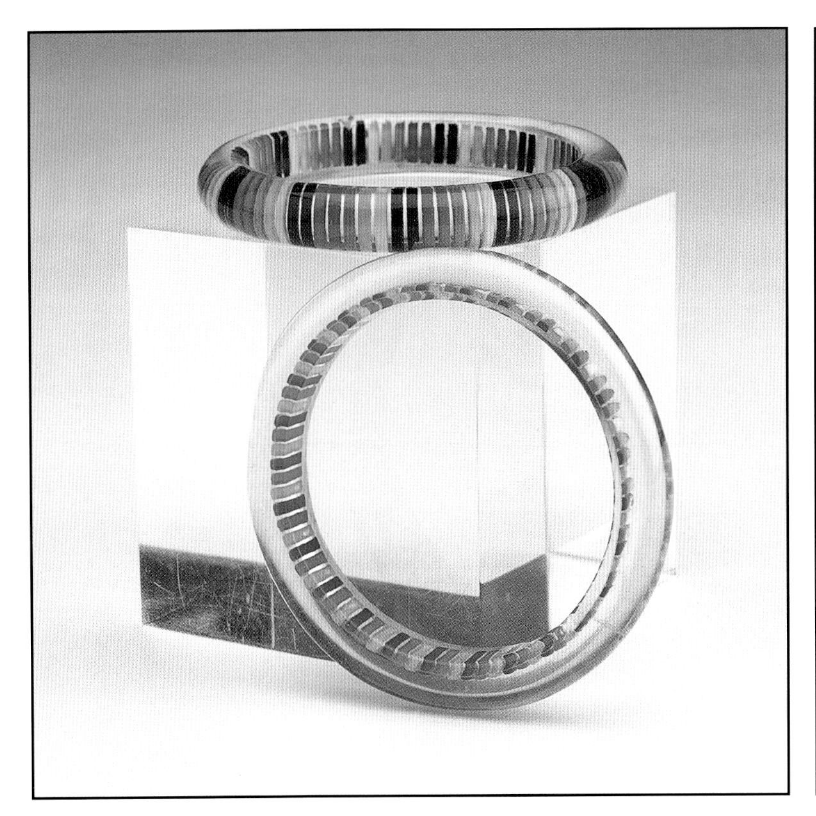

Colored stripes embedded in Lucite, $25-35.

Stack of blue and white Lucite bangles showing different techniques. From top, cast carved, $45; blue and white cast, $20; diagonal two-part glued bangle, $25; cast carved, $35.

Delicate Lucite open bangles with layer of metallic threads sandwiched in, $20.

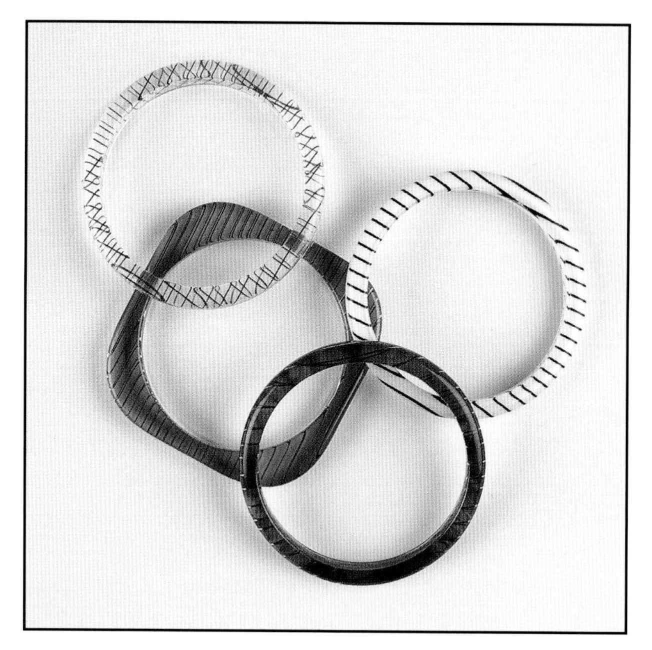

Layered laminates with metallic thread, round and square. $15-20.

Fabulous laminated bangles with clear inside layers, some cut on the bias, $15-35.

Set of green and purple Lucite bangles and beads, $25; pink and purple Lucite Best Plastics necklace and earrings, $35.

Stack of striped bangles and open bangle, $10-20.

Group of wide laminates, flat and saucer-shaped. Saucers courtesy of Bill Ruffin of the Bangle Barn. $25 and up.

Diagonal patterned bangles. The sliced ones are made in two parts and then glued together. The clear and white one is by R. J. Graziano. $20-45.

Pair of extravagant hard plastic saucer bangles. Each is made up of two pieces glued together. $20 each.

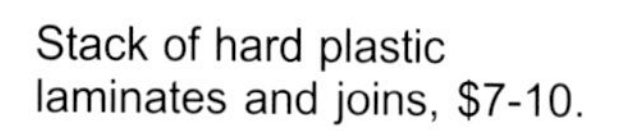

Stack of hard plastic laminates and joins, $7-10.

Lucite laminates with slinkies. Bangles, $15-25.

## Carved and Molded

Wide swirl-carved Lucite tortoise and cream bangles, resembling Bakelite. Tortoise, $85-100; cream, $45-55.

Cream Lucite or urea formaldehyde bangle with palm-tree carving highlighted with gold paint. This pattern was made in Bakelite in several colors. The cream was made at the same time, using another plastic to insure that it would keep its original color, $95.

Another pair of tortoise and cream carved bangles often mistaken for Bakelite. The cream may be urea formaldehyde. (Also found in black, with a matching necklace and earrings.) $30-35 each.

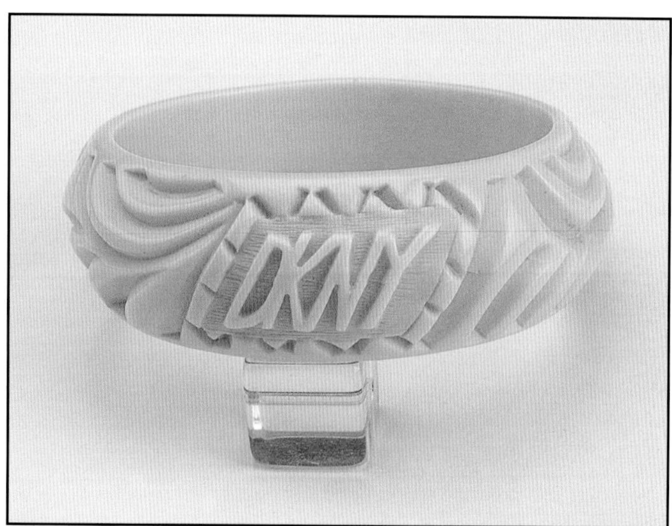


DKNY [Donna Karan New York] bangle in a pineapple and leaf pattern similar to vintage Bakelite patterns. The designer's logo is wittily worked into the design in relief. Also seen in black. Molded Lucite, probably 1980s, $35-50.

Group of molded white Lucite bangles, $10-20. Thumbprint carved bangle on top, $40.

Ivory-like carved Lucite bangles with "grain" to complete the effect. Wide, $35-40; narrow, $25-30; bottom, $20.

Group of Lucite and hard plastic bangles carved and painted. The white and blue one was made in France and is labeled "fait main" ("handmade"). $12-20.

Black cutback floral Lucite bangle, $6-8.

Lucite bangles with a colored outer layer, cut back to white or cream interior, $10-15; dots, $20.

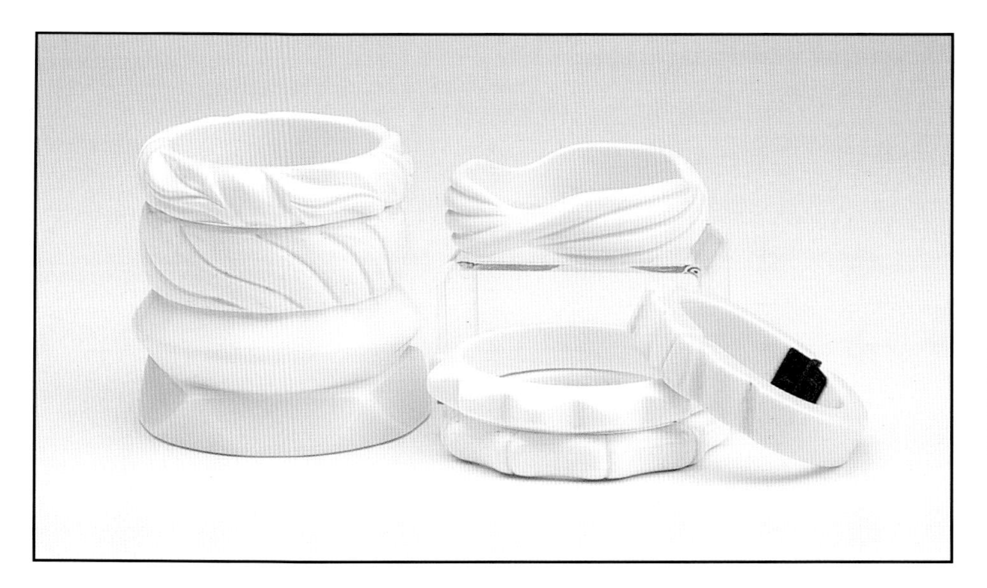

Group of cream or ivory colored molded Lucite bangles, $10-20. Hinge bracelet carved in a Bakelite pattern, $30-40.

These bangles were recently made from recycled soda bottles, $15; carved $20; carved with rhinestones, $30.

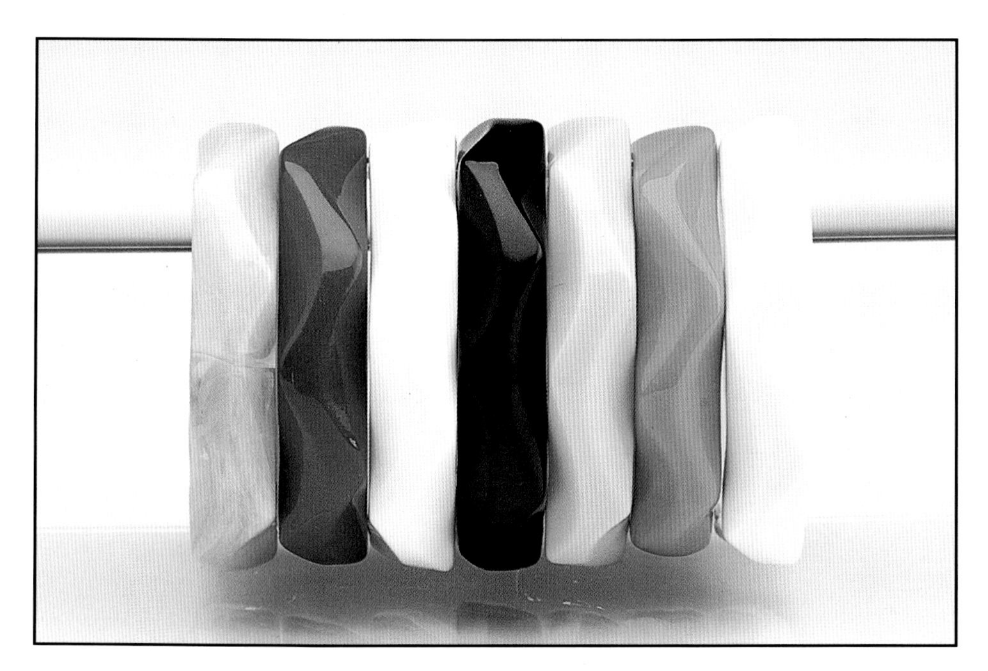

Six molded Lucite bangles in geometric pattern, $5-8.

## Ropes, Twists, and Facets

Bangles made of Lucite bent into spiral and knot-like shapes, $20-45.

Molded Lucite and hard plastic rope and twist bangles, $5-15.

Group of teal, tortoise, and black Lucite and hard plastic bangles in rope and twist shapes, $5-25.

Hard plastic facetted skinny bangles, $4-8.

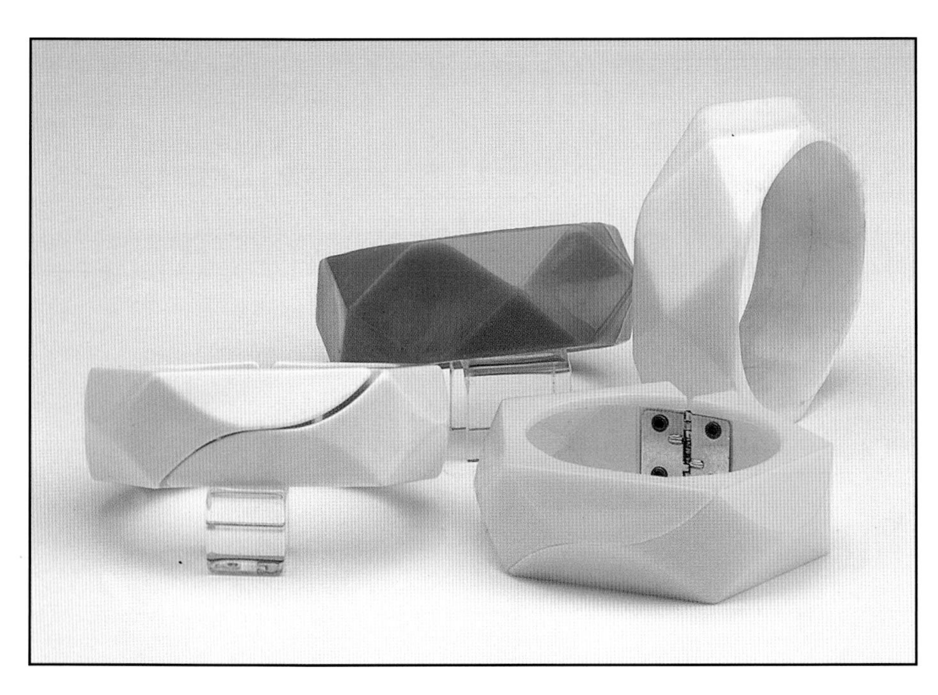

Hard plastic hinge bracelets and bangles in similar pattern. Hinges, $20; bangles, $10.

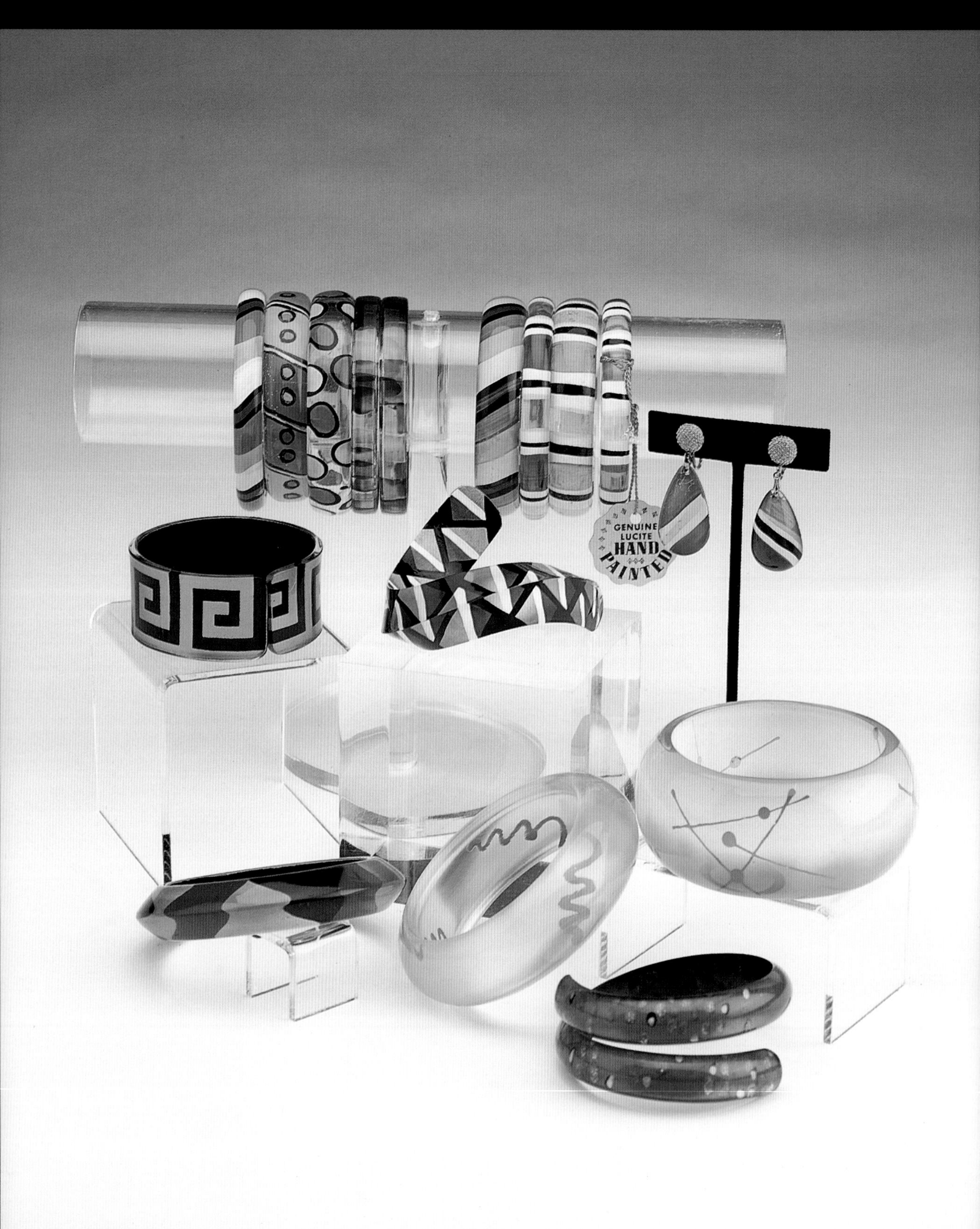


Group of reverse painted Lucite and hard plastic bangles. Top row, $10-15; green and cream set, $20; twist black, white, and grey, $25; wide cream with green, probably French, $35; rest $10-15.

Set of three reverse painted Lucite bangles in metallic colors, $25-40; lower right-hard corner, French Lucite bangle with colors fused inside, $40.

Painted Lucite and hard plastic bangles, Lucite cuffs, $20-25; stack of hard plastic bangles, $10; pearly white Lucite, $15; rest, $20-30.

Clear hard plastic and Lucite bangles painted with translucent animal print patterns, $15-20; striped cuff, $15.

Green hard plastic bangle with silver-leaf design, partly worn off, $35 and up if perfect.

Clear carved and reverse-carved Lucite bangles with Lucite and rhinestone purse. Carved, $35-45; reverse carved, $65-95.

Old carved and reverse-carved hinge bracelet, probably 1940s, $65-85.

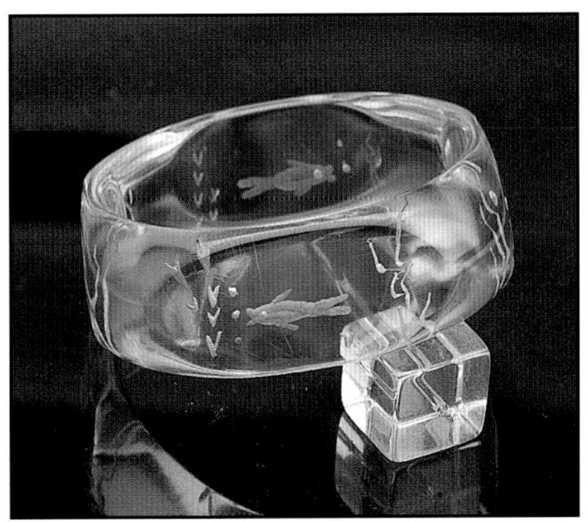

Reverse carved octagonal bangle, unsigned but authenticated early Shultz. The fish is a favorite motif of theirs. $100 and up.

New clear Lucite bangle, exuberantly carved in the spirit of earlier pieces, but unmistakably contemporary, Italy, $35-45.

Reverse-carved cuffs with stain or paint applied to highlight carving, $25-45.

Reverse painted and reverse carved and painted Lucite cuffs and open bangles, $35-55 and hard plastic bangles, $15-20.

Contrasting pair of reverse-carved painted Lucite bangle (left), $55 and Lucite bangle with inclusions, late 1950s or 1960s, $35-45.

## Inclusions

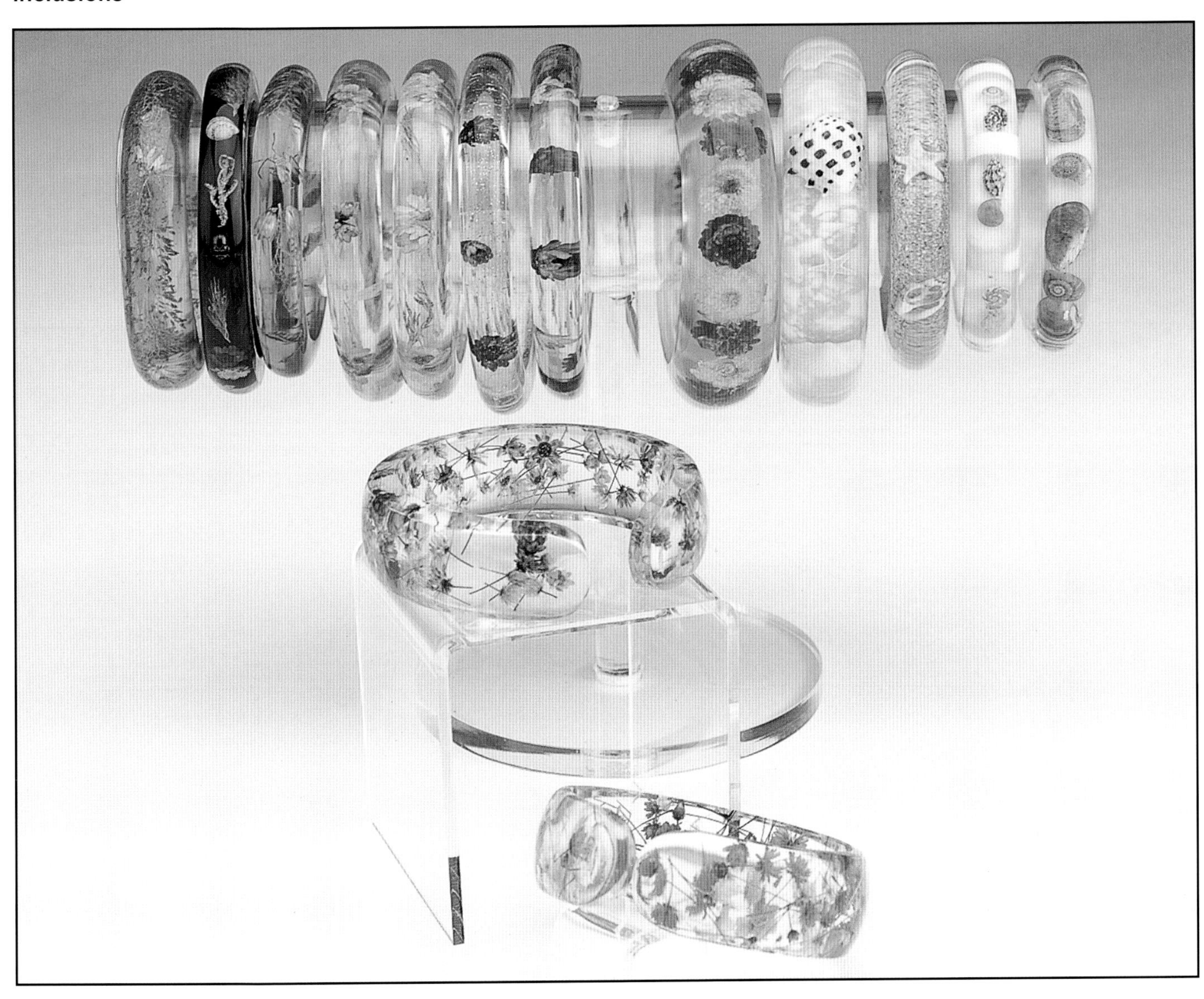

Inclusions of shells, flowers, insects, and leaves, $15-45.

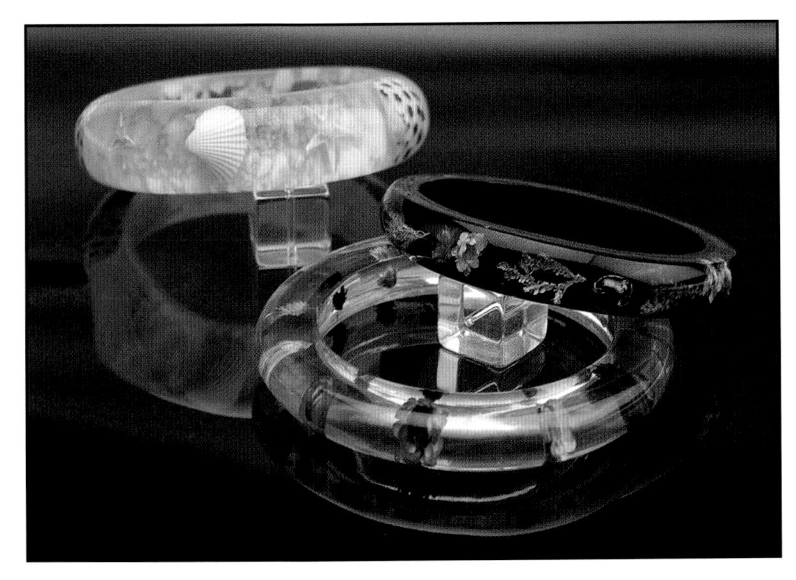

Three spectacular inclusion bangles with shells, leaves, and flowers, $20-45.

Jello-mold bangles: red and blue with mother-of-pearl inclusions and a green laminate, $25-35.

Shells, glitter, and opaque confetti bangles and hinge bracelets, with matching earrings. Hinge bracelets, $35-45; bangles, $15-35; earrings, $15-25.

An assemblage of glitter hinge bracelets and earrings. Hinge bracelets, $35-50; open bangles, $30; small button earrings, $15-20; larger earrings, $25-35.

Multicolored confetti hinge, bangle, and earrings. Hinge, $45-50; bangle, $35-40; spiral earrings, $45; button earrings, $20.

Blue confetti hinge bracelet, $45-50; earrings, $15-35.

Group of Lucite laminates with layer of glitter inside, $15-30.

Wide flat laminated glitter bangles with matching space-age earrings, $40-45 the set.

Group of laminated glitter bangles with matching fish earrings, bangles, $15-20; earrings, $15.

Pair of cabana striped fabric laminates, $35-40 each.

Group of cuffs and open bangles with fabric laminated between layers of Lucite. Floral cuffs, $15-20; cabana stripes, $35-40; gingham and plaid, $25-30. Sets with matching earrings, $35-45.

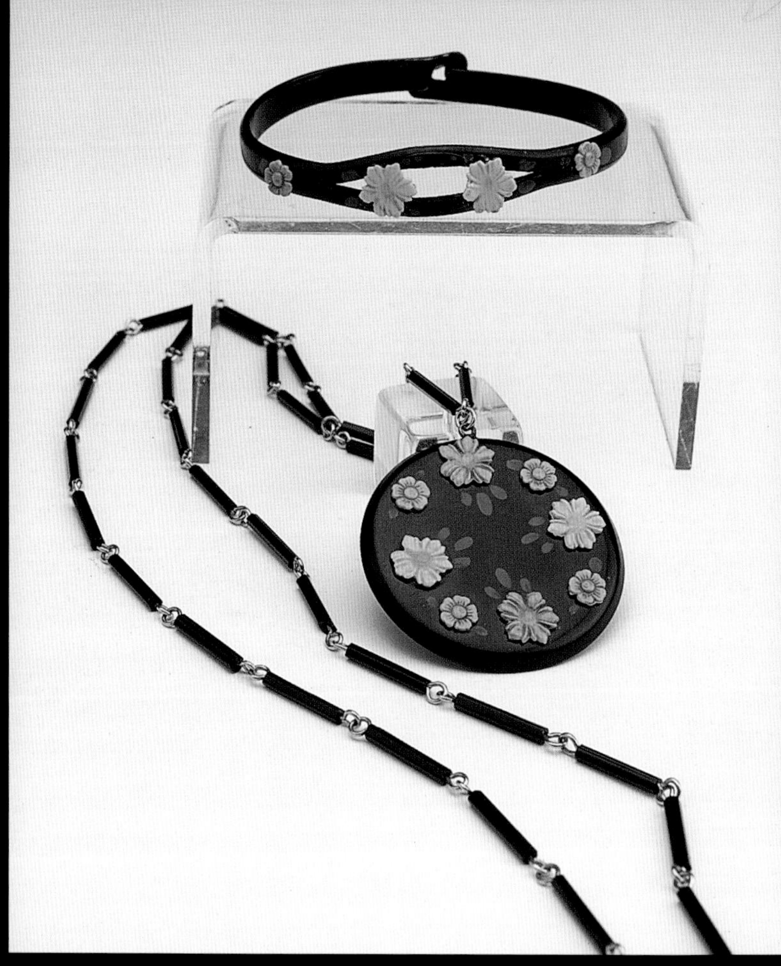

Celluloid bracelet and matching pendant with applied plastic flowers, $35-45.

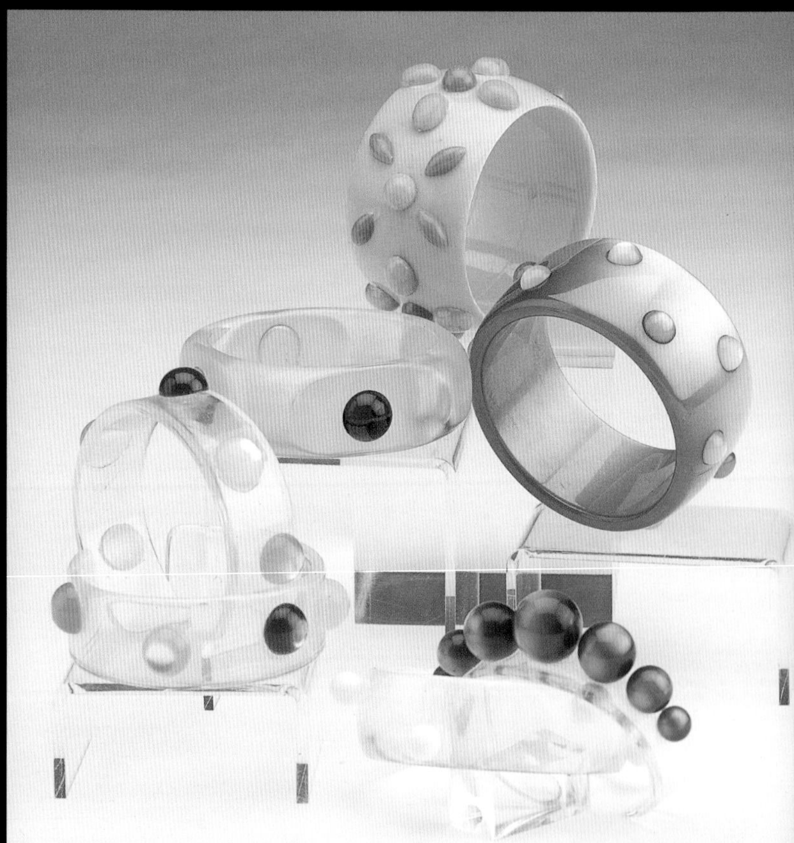

Bangles and cuffs with applied beads and cabochons. Clockwise from top: pink hard plastic bangle with cabs, $20-25; Italian blue moonglow bangle with moonglow cabs, $65; Lucite cuff with blue moonglow beads, $30; clear Lucite bangle with white beads, $25; pair of matching cuffs with moonglow cabs, $35 each; yellow Lucite octagon with glass cabs, $35. The Italian bangle is recently made. The rest are vintage pieces. Italian bangle, courtesy Lori Kizer.

Black Lucite stretchy bracelet with applied plastic strawberries strung on silver elastic, quite small, $25-35.

Group of Lucite bracelets with metal and plastic appliqués including coins, stars, opal glass cabs, and checked or marbled dots, $20-40.

Hard plastic hinge bracelet with applied cloth flowers – quintessential fifties and very hard to find, $30-40.

Plastic bangles with metal elements. Clockwise from top, heavy Lucite with magnetic hinge, $25-30; metal bangle with moonglow "chiclets," $15-20; celluloid tortoiseshell cuff with metal trim, $25-30; silver-tone bangle with Bakelite ends, $20; golden marbled Lucite with gold-tone X's, $15; pair of hoop earrings, $10.

## Rhinestones

Celluloid "sparkle" bangles, with two bypass bangles, one with rhinestones, $35-85.

Lucite and hard plastic with applied multicolored rhinestones. From top, clockwise: black set with turquoise rhinestones, $45 and up; shaded red to pink bangle and earring set, $35-45; white Lucite bangle with plastic foil-backed cabs, $30; dark green Lucite with green and clear stones, $45-60; wide black hard plastic with foil-backed plastic cabs, $15-20; purple celluloid dangle hoop earrings with stones, $18-20; cranberry Lucite with stones, $25-35.

A group of Weiss and similar rhinestone hinge bracelets with earrings, often taken for Bakelite, but actually made of Lucite and other plastics. Hinge bracelets alone, $65-85; sets with earrings, $125-150.

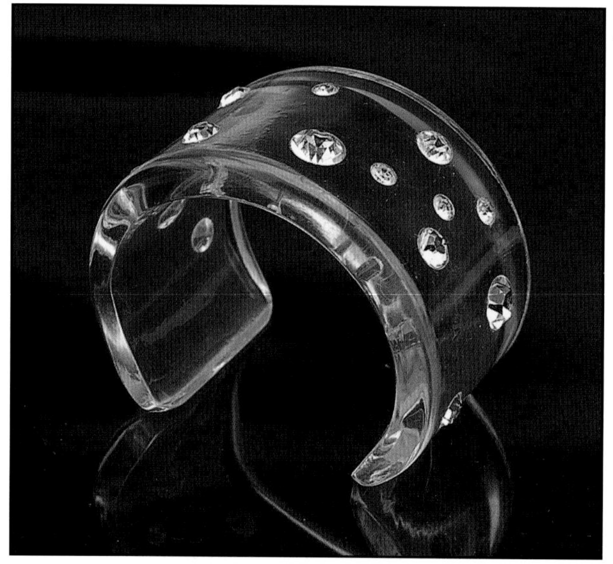

Chunky Lucite cuff with rhinestones, $35.

Black Bakelite hinge bracelet with square-cut stones, $125.

Rhinestone Lucite bangles and two wrap bracelets, $25-45.

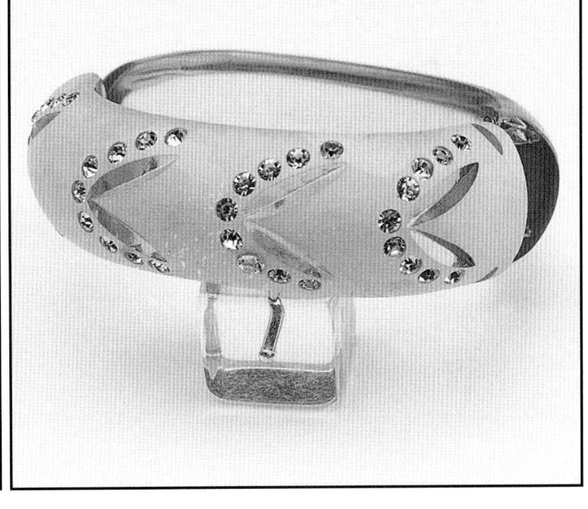

Leru transparent Lucite hinge bracelet with rhinestones. The front is coated and cut through to show the transparence. $50-60.

Group of Leru Lucite and celluloid with rhinestone hinge bracelets and earrings. Set with earrings $75-100; others $50-75.

Weiss-type unsigned hinge bracelet in cream with multi-colored rhinestones, and co-ordinating pierced earrings, $65 and up.

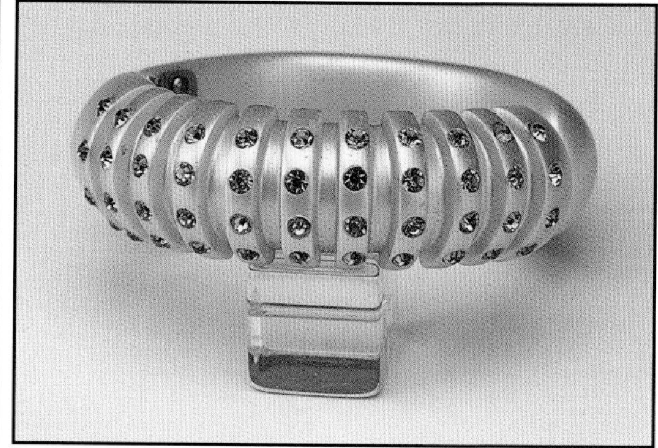

Pink iridescent coated hinge with pink rhinestones. In the spirit of Weiss, but of cheaper quality, $35-40.

**Moonglow**

Moonglow Lucite bangles, $5-25.

Stack of narrow jewel-tone moonglow Lucite bangles, $5-8 each.

Substantial dark blue moonglow stretchy bracelet, $25.

Pearlescent bypass and spiral bangles with green hinge bracelet, $25-40.

Mother-of-pearl patterned moonglow bangles, $20-40.

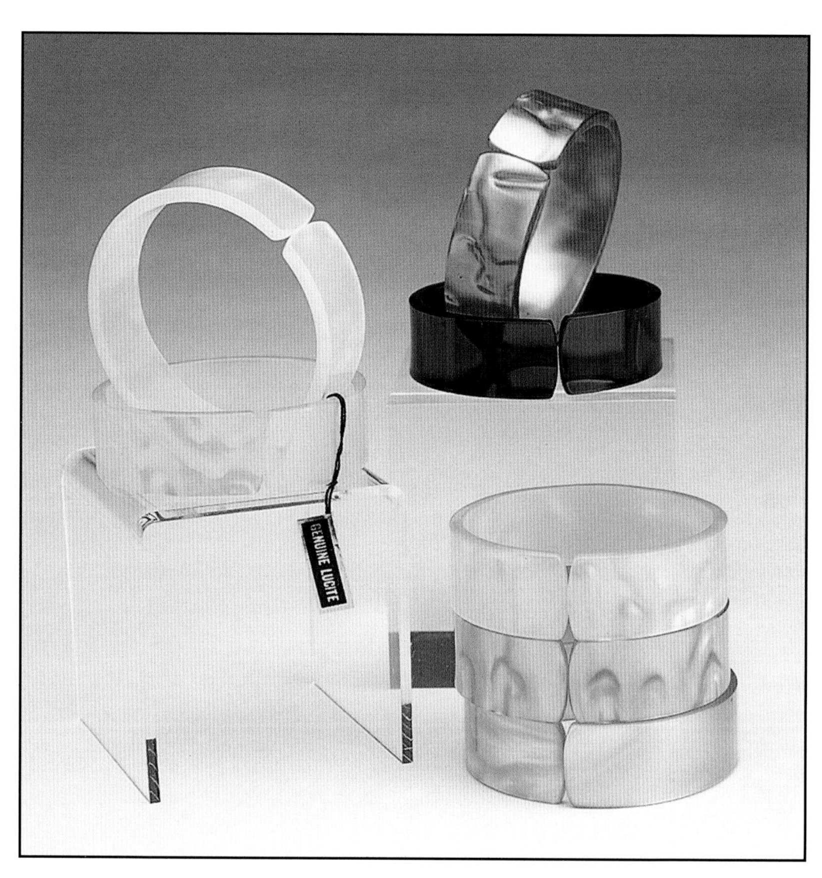


Sliced Lucite open bangles, probably 1990s. The pearly pattern is especially marked in this type, $7 each.

A group of unusual variants on moonglow, including two wide open bangles ($35), a group of mother-of-pearl patterned bangles ($15-25), a matched pair of blue tapered bangles that together look like a single bangle ($20), and two squared bangles ($30 pair).

Two recently made moonglow bangles on top of a blue vintage one. Note the higher contrast, sharp edge, and chalky appearance of the inside on the new bangles. New, $5; vintage, $8-10.

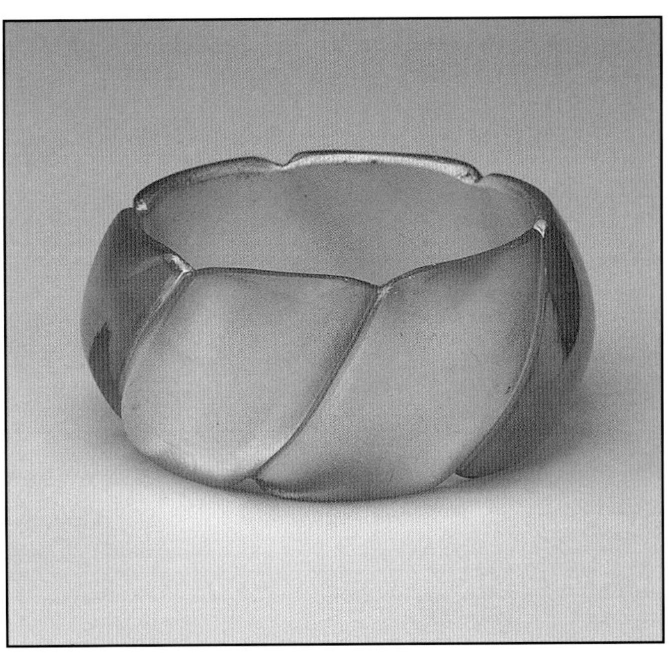

Impressive Lucite bangle with lobe carving. Appears to be moonglow at first glance, but it is actually clear with metallic paint inside, $30-35.

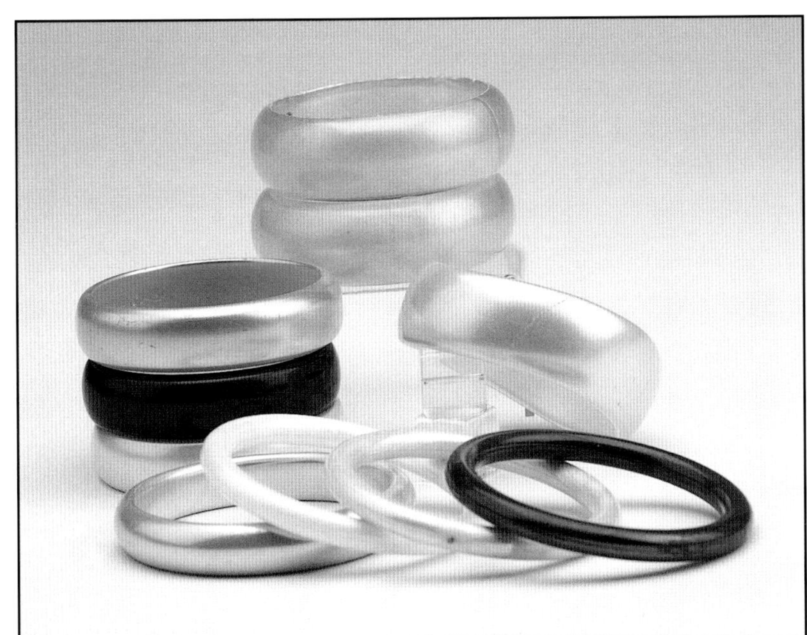

A group of hard plastic pearly bangles. These are sometimes confused with Lucite moonglow, but the pearlescence is in the coating, while moonglow is made by mixing powdered shell into the plastic itself, $4-8.

Moonglow bangles, carved and molded. The patterns on the molded ones (blue, center and pink, right) are more regular and shallower. Molded $15; carved $25-30.

Group of unusual carved moonglows, $25-75.

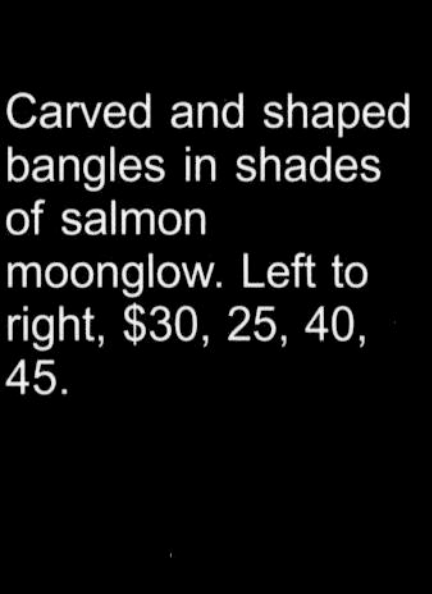

Carved and shaped bangles in shades of salmon moonglow. Left to right, $30, 25, 40, 45.

Squared laminated bangles in two tones of moonglow, $20-40.

Moonglow bangles in metallic shades, including laminates and two slant-cut bangles in foreground, $20-40.

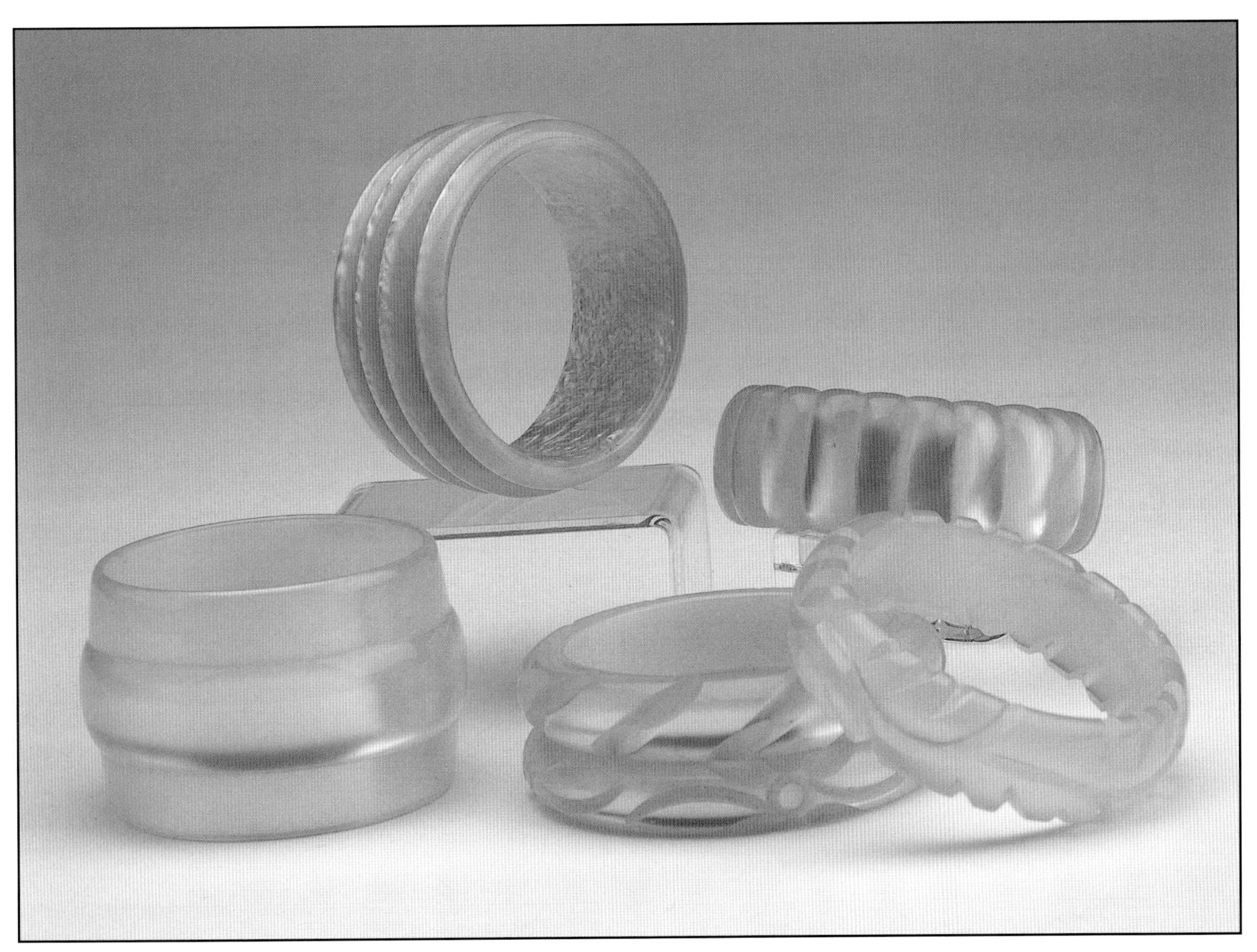

Group of white and ivory carved and shaped moonglow bangles. $20-45. Especially fine floral carved on right, $40-45.

Layered and laminated moonglows, $10-35.

Stack of unusual irregularly striped moonglow bangles, $20-35.

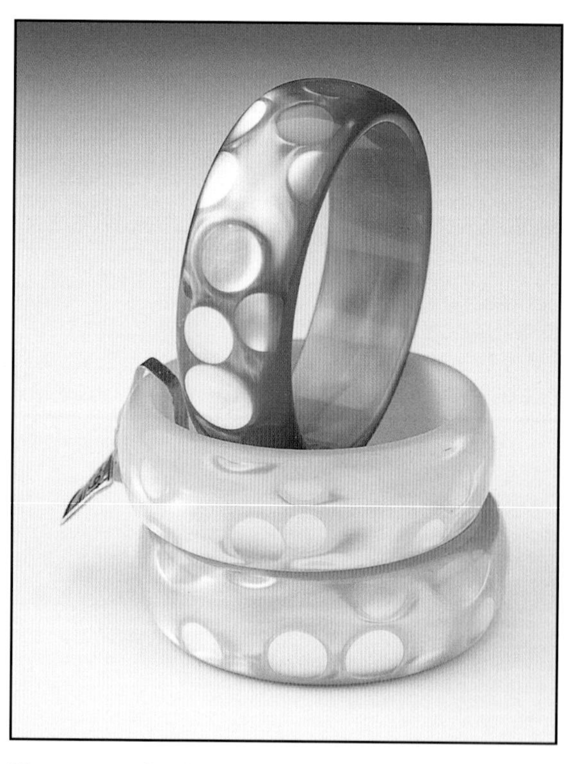

Three moonglow bangles with inset pearly dots, c.1963. Tag on blue bangle reads "Festival Design." Hard to find, $35-45.

Lavender pearlescent cuff with an array of moonglow and pearlescent earrings showing different techniques, $10-25.

Mountains of Lucite beaded necklaces, bracelets, matching pins, earrings, and a large ring.

Different kinds of moonglow bracelets, including stretchies and memory wire bracelets, some wrap-around, and some with dangling beads, $15-40

Set of pale blue moonglow memory-wire bracelet, with label "Satinore by Richelieu," and matching necklace and earrings; pin, $20; clip $35.

Cranberry moonglow memory-wire bracelet with necklaces, earrings, pins, and rings, $10-40. Darker set of dangles on coated chain with matching earrings, $55.

Pink barrel-shaped moonglow pop-it beads, $18.

Very unusual open bypass bangle in dark blue moonglow, $35 and up.

Mother-of-pearl Lucite bypass bangles in unusually intense colors, $35-$45. "Matching" Lucite napkin rings $3-5 each.

Iridescent Lucite bypass bangles. The pearlescence has less depth than most moonglow, $20-25.

Stacking and interlocking sets of bangles, clockwise from top: Cream, pink, and green Lucite zigzag stacker, $25-30; two pair of green Lucite fit-together bangles signed "E.S.K." $15-20 each set; red, white, and blue hard plastic zigzag stacker, $15-20; stepped fit-together red, white, and blue Bakelite set (cream was originally white), $150 and up.

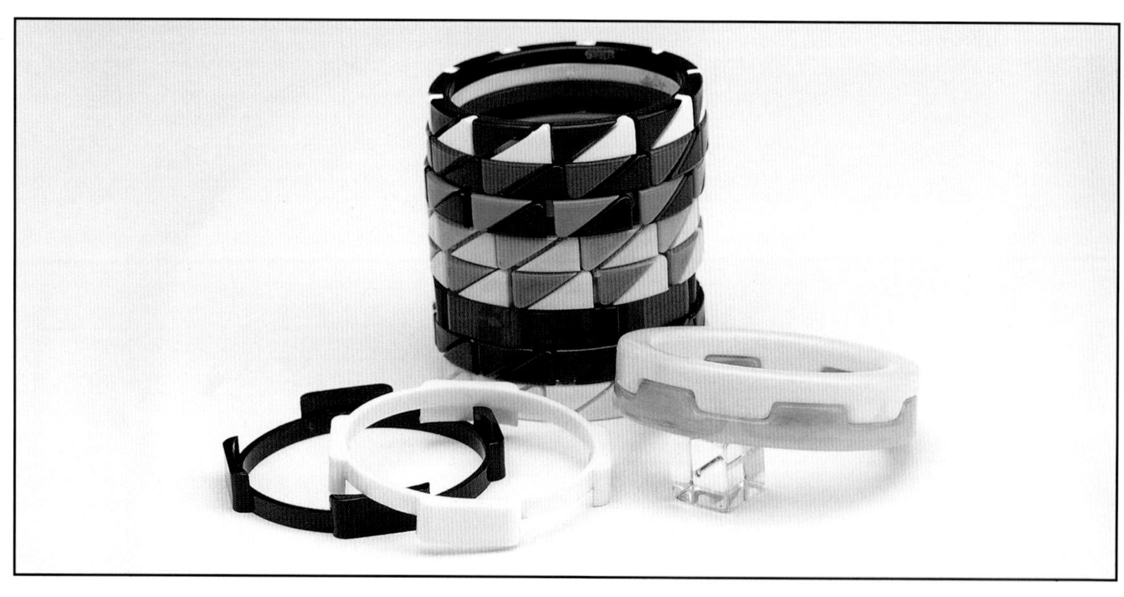

These hard plastic "puzzle" bracelets are made up of two parts that slide apart and can be recombined in different ways. Made in Hong Kong. Tan and cream hard plastic bangle is made up of two pieces glued together. $5-10 each.

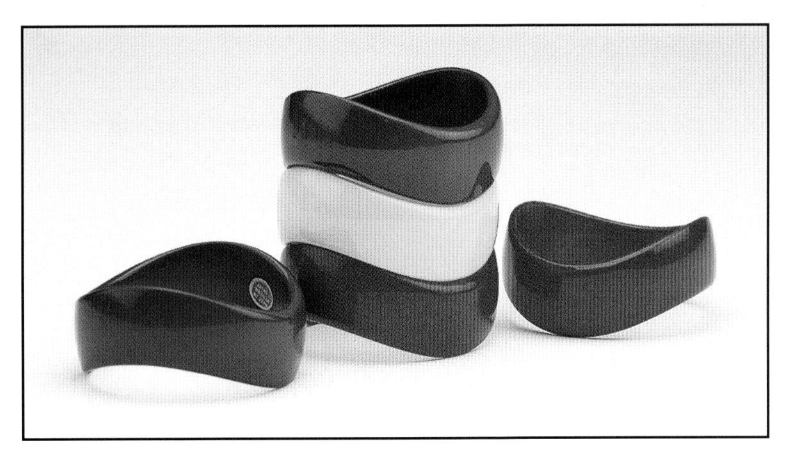

Soft plastic wave-shaped bangles, labeled "Made in Western Germany," $5 each.

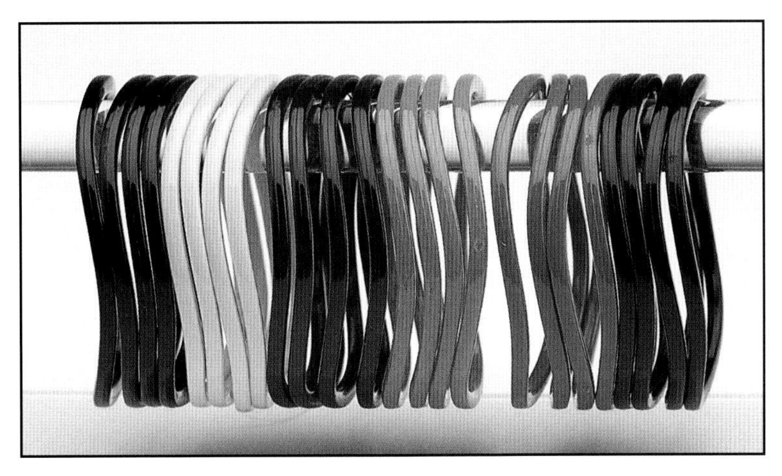

Set of hard plastic wavy bangles in earthtones, $10-15.

Wavy hard plastic bangles, some with dots $3-5. Matching black and white dangle earrings, $15-20. Frosted and pearlescent bangles are Lucite, $4-6.

Black and white or clear frosted square bangles, beveled squares, $30-35; two-piece magnetic bangle, $35-40.

# Identifications

## Avon Calling

Avon bangles, stretchies, and matching pieces, dated between 1977 and 1989.

Avon "Luminesque" Lucite bangles with box, dated 1977. Also found in white, black, and cranberry red, $15-20.

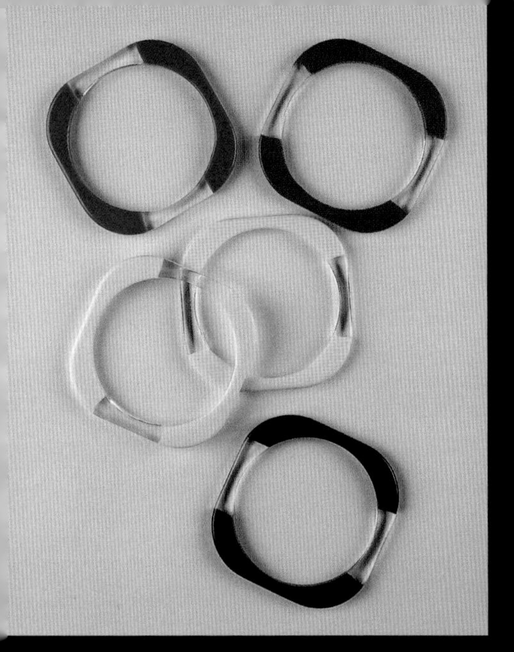

Avon set of Lucite “Summer Rainbow” bangles, 1978, $25-35 set.

Avon “Color-Go-Round” bangles. The dates on the boxes are 1977 and 1979. The ones shown here test positive for Bakelite, $15-18. Also found in pastel solid colors in Lucite, $6-10.

Red dotted white hard plastic Avon bangle, 1987, with matching earrings, $12-14; bangle alone,

Passion Flower design, 1989 and Tropical design, late 1980s, hard plastic bangles with decal decoration. Like many Avon bracelets, these came in two

## Signed Pieces

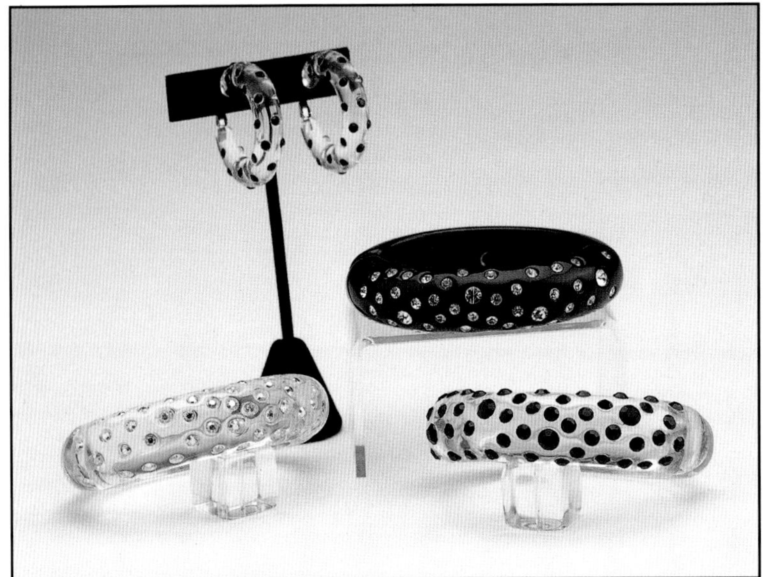

Hinge bracelets by Kenneth Jay Lane, whose award-winning designs were worn by Jackie Onassis, Elizabeth Taylor, Audrey Hepburn, and Princess Diana. Bracelets are signed "KJL." Hinge bracelet with matching earrings, black with clear stones, $100-150; bracelets, clear with aurora borealis stones, and black with clear stones, $75-100.

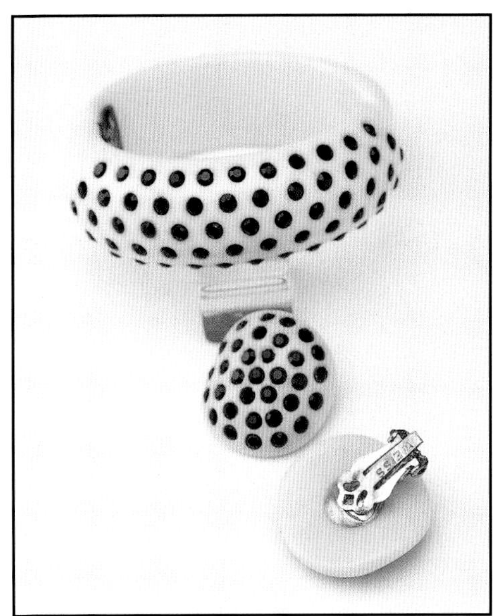

Signed Weiss Lucite set, cream hinge bracelet and earrings with black stones, $125.

Group of signed Monet Lucite bangles, most from the 1980s Directives series. The transparent yellow and rust "waist" bangles are unsigned and may not be Monet, although they are shaped exactly like the marbled rust bangle with Monet plaque ($30). Directives series, $35-45; cuff and set of three wavy bangles (all with gold Monet plaque), $20.

Two Monet Directives bangles with paper label, showing price of $25. Current value, $35-45.

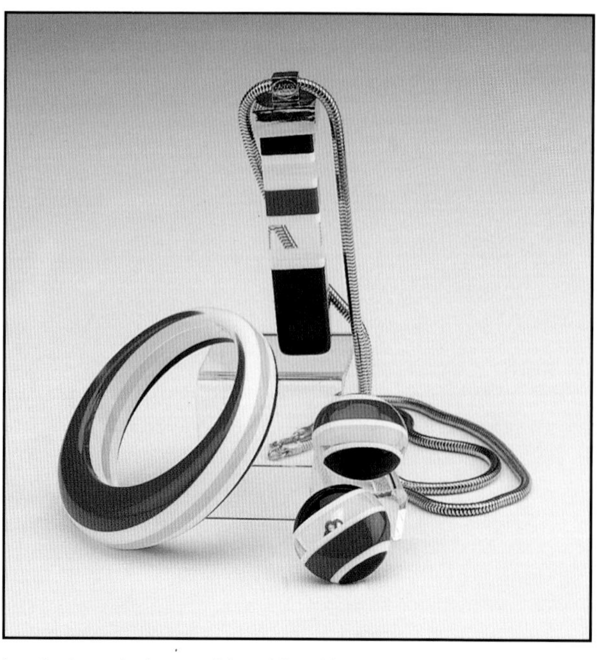

Lucite bangle layered in white, blue, red, and clear Lucite with matching button earrings, and pendant signed "Lanvin Paris." Set, $75-100. A matching laminated square bangle can be seen on page 139.

Chunky clear Lucite cuff marked "E.L." This was a gift to employees of Estée Lauder participating in the introduction of the fragrance "Knowing." Team leaders were given the same cuff with diamonds set into it. As shown, $35-45.

Givenchy Lucite bangles with signature on gold-colored plaque, $15-30.

Black and white swirl Lucite bangle with "dvf" [Diane von Furstenberg] plaque, made by Best Plastics Company of Rhode Island (see chapter 6). While many of these were made, only a few seem to have been signed. With plaque, $35-40.

Stack of Sarah Coventry injection-molded rose bangles, most likely 1970s. Often mistaken for more expensive Lucite or even Bakelite, these hard plastic bangles give themselves away by the visible seams and an obvious hollowness when tapped. Only one of these still has its signature "Sarah," which was used by the Sarah Coventry after 1952. All cream, and cream with tortoise are quite common. The other colors are extremely hard to find. The cream and green one is labeled "W. Germany." $12-15 for common colors; up to $30 for red.

### European Plastics

Pyramid of bangles with a retro art deco design. This extremely popular style was made in an astonishing range of colors, not all of which are shown here. These are molded, not carved, as is shown by a pair of seams running opposite each other through the middle of the design on the outside of the bangle. These seams are more apparent on some bangles than others, possibly reflecting changes in the place of manufacture. These were originally made in Germany probably during the 1960s but later production moved to Hong Kong. The magenta and fuchsia ones have seams along their edges and are hard plastic, $7-12 depending on color; transparent colors $15-20.

"Deco" style bangle with three different labels. The earlier pieces have labels reading "Made in Western Germany." Later pieces are labeled "Genuine Lucite Made in Hong Kong" or merely "Genuine Lucite" without country of origin.

Pale frosted leaf-patterned West German Lucite bangles, $15-$20 for set.

Group of floral molded Lucite West German bangles, $7-10. Translucent tortoise- shell with higher relief, $12-18.

Heavy tortoise Lucite open bangle and lighter aqua spiral bangle both with molded raised signature "Design JO W. Germany." These have a pebbly finish on the inside surface. $20, $15.

Aqua transparent Lucite spiral bangle showing a molding mark or sprue, which resembles a pontil mark like those found in blown glass.

Enormous spiral molded bangles, also seen in other colors including frosted clear. The transparent ones are Lucite, while the opaque are softer and possibly made from another plastic. Label: "Made in Western Germany." Opaque, $15-30; transparent, $35 and up.

West German Lucite bangles with deep swooping grooves. Opaque $8-12; transparent $12-18.

Hard plastic West German bangles. Swoops $5-8; beaded rings $8-12.

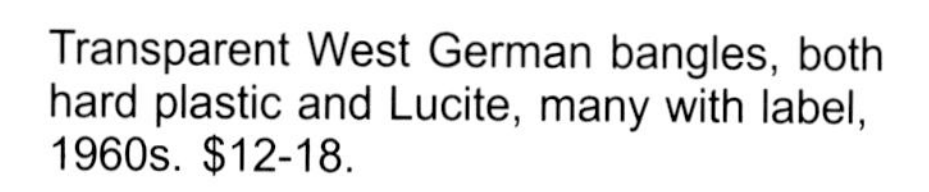

Transparent West German bangles, both hard plastic and Lucite, many with label, 1960s. $12-18.

Close-up of pink and blue transparent West German bangles. The pink one, second from left, shows fine "drips" often seen on transparent German Lucites. They are traces of the manufacturing process and, while they do not enhance the value, they are not considered damage. Lucite bangles, $12-15. The two outer bangles are hard plastic, $8-10.

Clear cuff and open bangles with central stripe of opaque color. The orange one is labeled made in "Western Germany." $10-15.

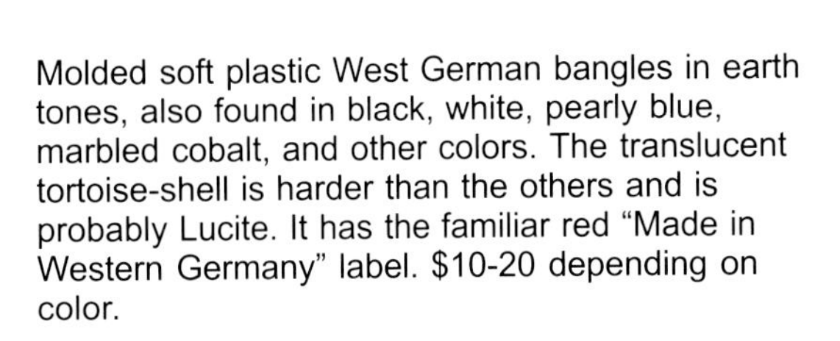

Molded soft plastic West German bangles in earth tones, also found in black, white, pearly blue, marbled cobalt, and other colors. The translucent tortoise-shell is harder than the others and is probably Lucite. It has the familiar red "Made in Western Germany" label. $10-20 depending on color.

Buch + Deichmann of Denmark brought out these vintage-inspired pieces in the 1980s in a plastic that is slightly softer than Lucite. The wavy bangles and the stylized florals are dated 1981. The open geometric bangles are signed but not dated, while the narrow tube bangles and the box for the "Combi" set are signed with the name of the designer, Ketty Dalsgaard. Note the elegant design of the folding brush ($20). Medium bangles $15-20, narrow ones $8-12, set of spacers in box $20; pins $12-20; pill boxes $14.

This close-up of a B + D bangle dated 1981 and a similar West German one raises some interesting questions, since the West German one should by rights be about 20 years older than the Danish one. Was B + D imitating post-war imitations of pre-war pieces?

Group of bangles all marked "Made in Italy." The colorful opaque bangles are made of two halves that fit together, 70s or 80s, set $30. The red, white, and black pair is probably from the early 1960s, $20. The transparent flying saucers with a central opaque stripe are somewhat newer, set $30-40.

Lucite domed gold glitter bangle made in France in the 1960s, $50.

French Lucite bangles with metallic threads inside, 1950s. The larger pair with copper threads has a translucent purple coating on the inside that gives it a particular richness. Narrow bangles, $35 a pair; wider bangles, $35 each.

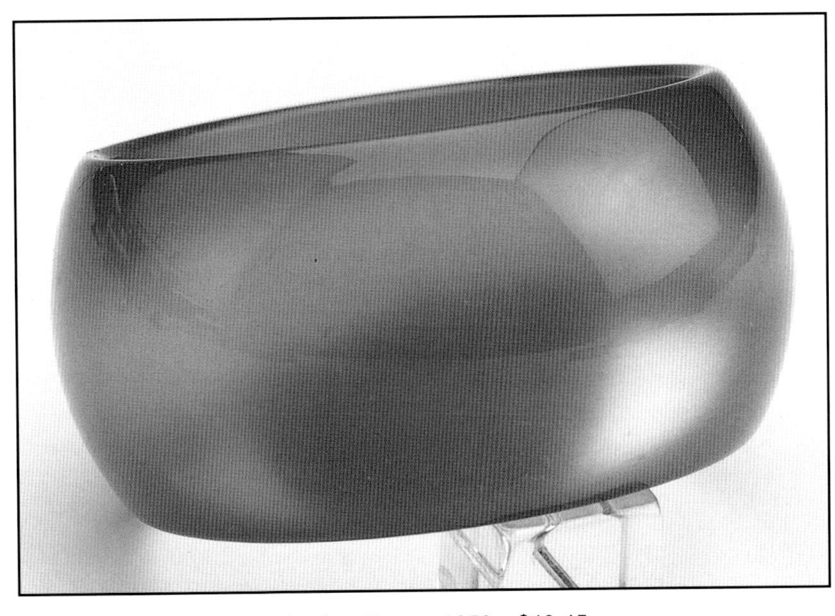

Luminous lilac moonglow Lucite, France 1950s, $40-45.

Clear Lucite with embedded black and white petal-like confetti – two thick flat bangles, one chunky domed bangle, and matching earrings. Clear printed label "Marc Labat Paris." Probably 1970s. Bangles $45, 55; earrings $35-40.

Transparent Labat bangles suffused with vibrant color, signed with label, $35-45.

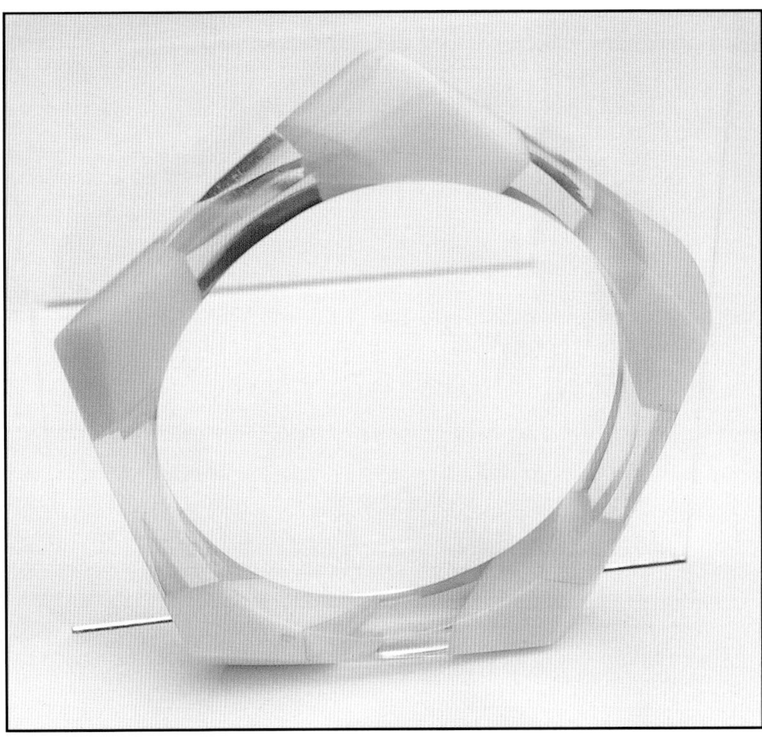

Elegant pentagon bangle in clear and white, France, 1960s, $35.

Lucite bangles, France, 1960s. Wide chunky bangle with golden dust, $60; multicolored pearly bangles, $40-50 a pair.

French bronze carved moonglow Lucite bangle, 1960s, $40.

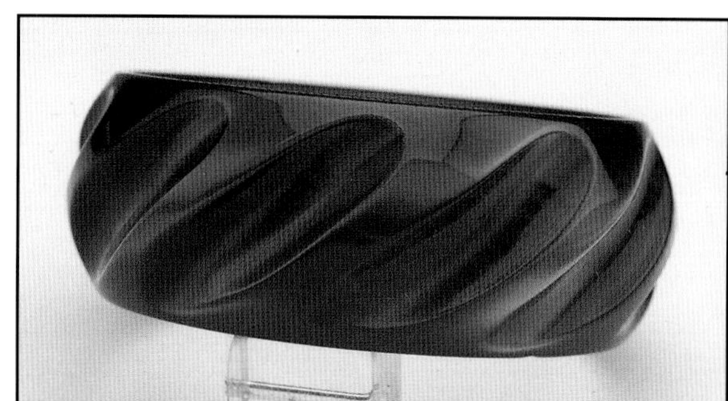

Carved lilac moonglow octagon bangle, purchased in London, but probably French, 1950s, $75.

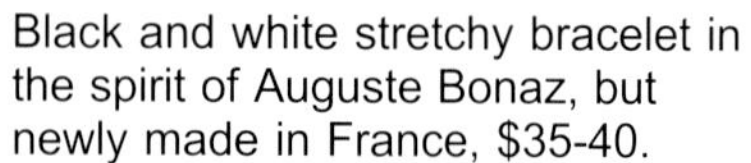

Black and white stretchy bracelet in the spirit of Auguste Bonaz, but newly made in France, $35-40.

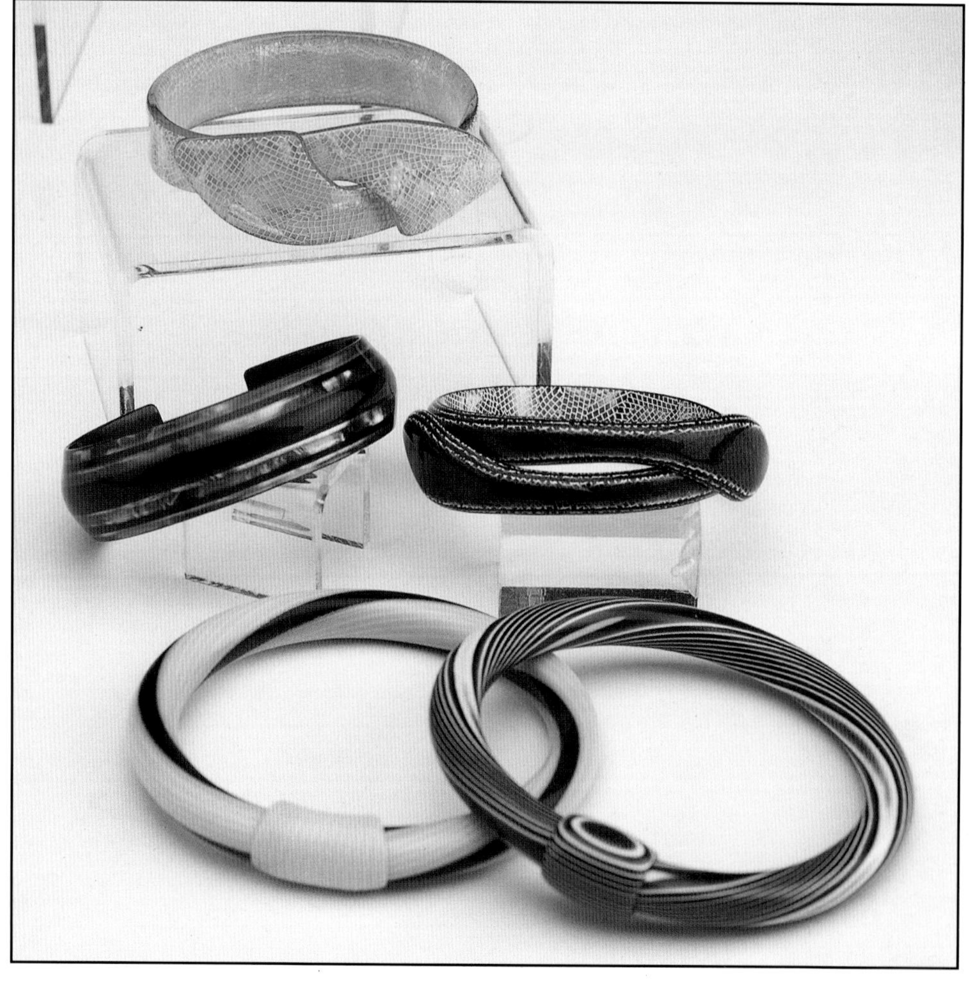

Group of Lea Stein bangles. Clockwise from top: pearly golden with delicate snake pattern in white sandwiched over transparent yellow, $75-85; red-brown pearly bypass bangle with seven layers and similar snake pattern, $45-50; candy-cane colored "knot" or "bead" bangle, $60-70; classic "bead" bangle showing detailed texture in the cream sections, $60-70; pearlescent red and pinky lavender cuff with at least five layers. $75-85.

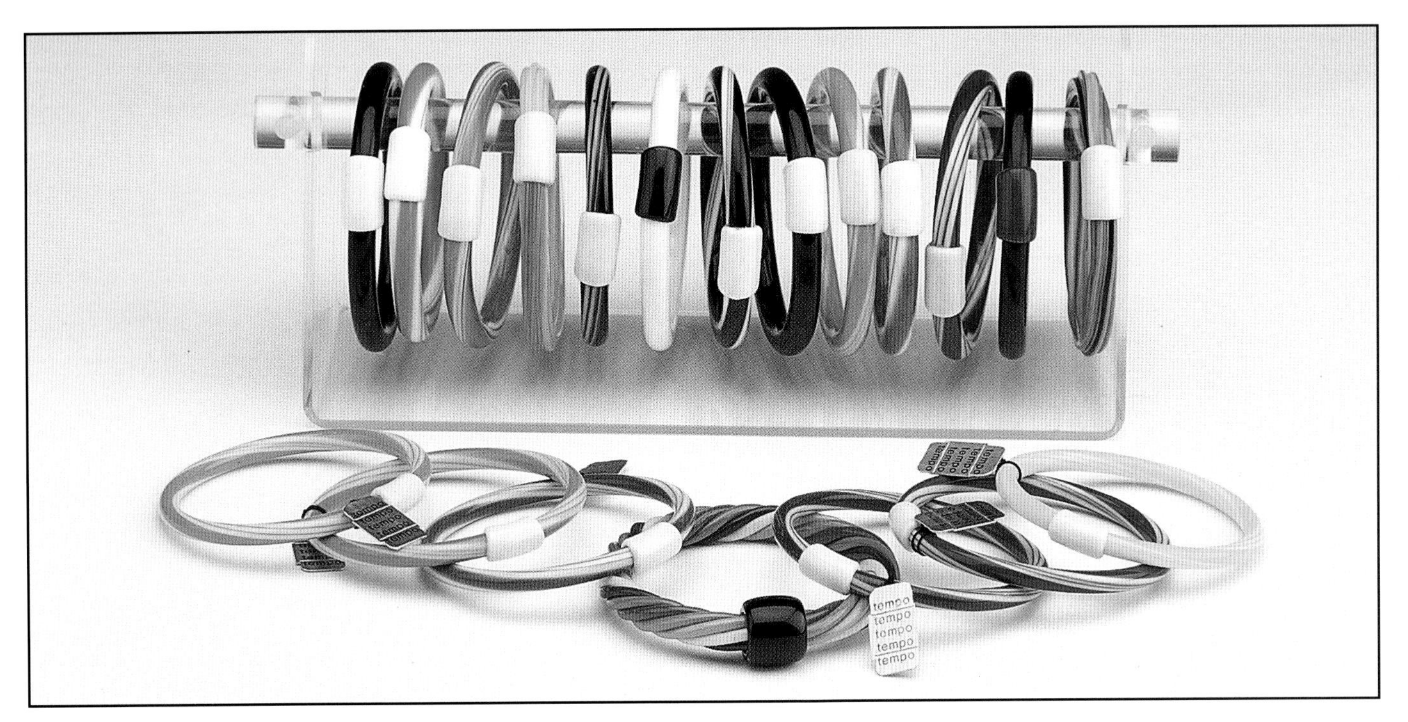


These knock-offs of Lea Stein's most recognizable bracelet were made by Best Plastics of Rhode Island. The smaller ones (foreground) were marketed under the name "Tempo," $10-15; larger ones (hanging), $10-25. (The brown and pink ones have raised stripes and were probably made elsewhere.) In the center foreground is a more elaborate version of unknown origin with different colored strands twisted together, also seen in a red, white, and blue version, $20-25

Enormous Lucite bangles by the Parisian designer Mme. Casalta in the 1960s. They are flat on one side and tapered on the other. Shown with an even larger and heavier red domed bangle, probably American. The term "runway" piece is over-used, but these are the genuine article. $50-75.

Pair of very large chunky transparent spiral bypass bangles with a triangular cross-section, also seen in dark red. $45-60

Watercolor transparents and translucents, $35-60.

Three unusually shaped and colored Lucite bangles, all shading from one color to another.
The rectangular bangle has a paper label with the name "Barkhor." $50-60; the others, $40-50.

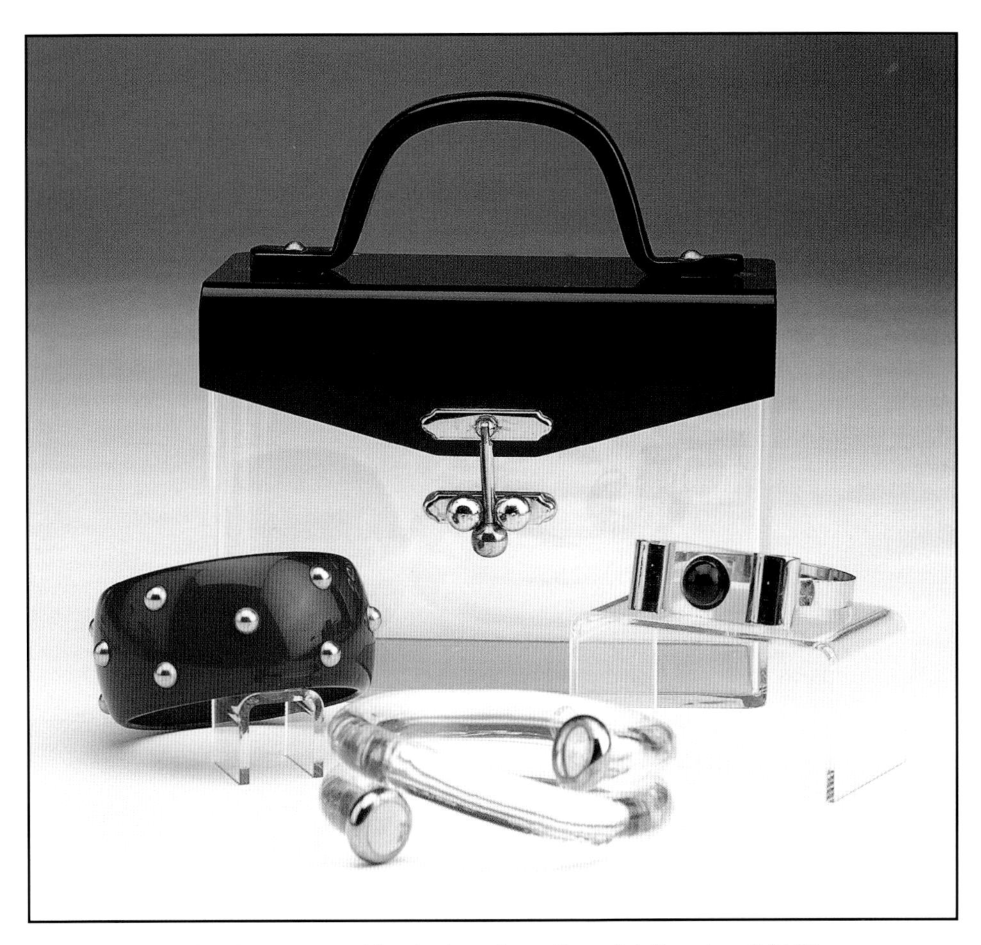

Black and clear Lucite purse and Lucite bangles with metal. Bangles, $45-60.

Artist-signed and dated "jmc 85" tapered Lucite bangle, black and clear with iridescent glitter, $75-95.

Lavender and gold-washed Lucite wavy-edged "saucer" bangles, $20-25 each.

Fabulous green and white laminated Lucite bead set of bracelet, necklace, and earrings, with similar earrings in pink and lavender. The necklace has green Bakelite beads interspersed. Set, $100; pink earrings, $15-25, large green, white, and clear button earrings with French style clips, $45.

Unusual Lucite teal knot bangle, $40-45.

Pink and cranberry Estée Lauder chunky sliced Lucite bangles, part of a recent make-up promotion, $8-10. Unusually shaped yellow Lucite bangle signed "Archaos Odile" along its frosted edge, 1980s, $35-45.

Huge green Lucite open bangle in the shape of a curved leaf, $40-45; neon pink wrap bangle, made up of approximately one yard of Lucite, $55; blue-green wrap bracelet, $30.

Delightful hard plastic transparent bangles that are actually functional tambourines! Probably 1960s, $15-20 each.

Pearlescent Lucite purse with pearlescent and rhinestone hinge bracelet; foreground, two unsigned Leru hinge bracelets, one with cabs, the other with matching earrings. Rhinestone pearly hinge, $45-55; plain Leru with earrings, $40; Leru with cabochons, $50 and up.

Chunky olive transparent Lucite bangle with sliced edges, with a clear Lucite bangle, $35-45.

Amazing magenta and black Lucite set with huge chunky square bangle, enormous dangling circle earrings, and leaf-design necklace, probably 1980s, $100 and up.

Lucite and sterling bangle, square with concave corners, $75 and up.

Clear, white, and colored geometric laminates. The red, white, blue, and clear one is by Lanvin and matches the set on page 118. $20-35.

Pale cognac-colored heavy Lucite bangle with carved rounded triangular facets, $55 and up.

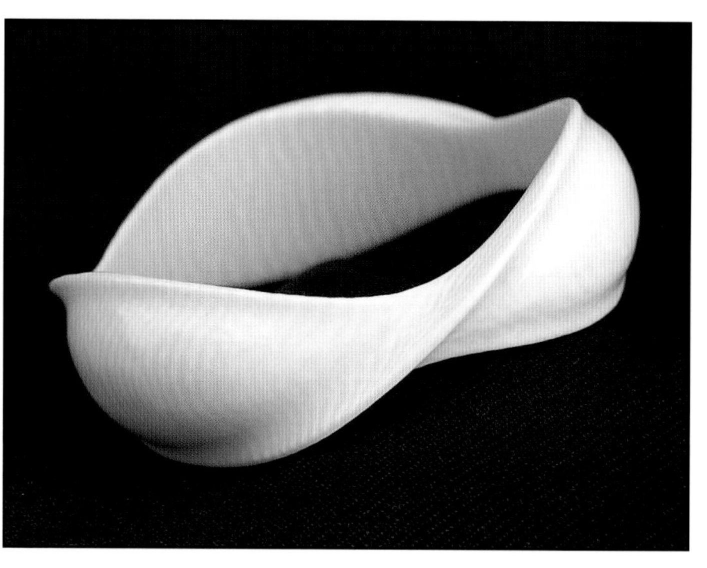

Imitation ivory bangle with grain markings, in an unusually delicate twist shape, $35-45.

Group of tortoise-shell Lucite bangles, with cream and tortoise laminate tube bangle, $20-45.

Teal Lucite bangle and clip with chrome accents, $85 and up.

Cellulose acetate triangular bangles with matching earrings, $45-60 the set; three unusual wavy round bangles embossed with crisscrossing lines, $55-65 the set; twisted bypass bangle, $20-25.

Warm golden Lucite bangles, $20-45.

Clear Lucite bracelets in a variety of types, $20-35.

Clear and deep blue transparent octagonal Lucites with beveled edges, $10-20 each.

Extraordinary bent Lucite bracelet with similarly shaped napkin rings. Bracelet, $45-55; napkin rings, $30 for set of six.

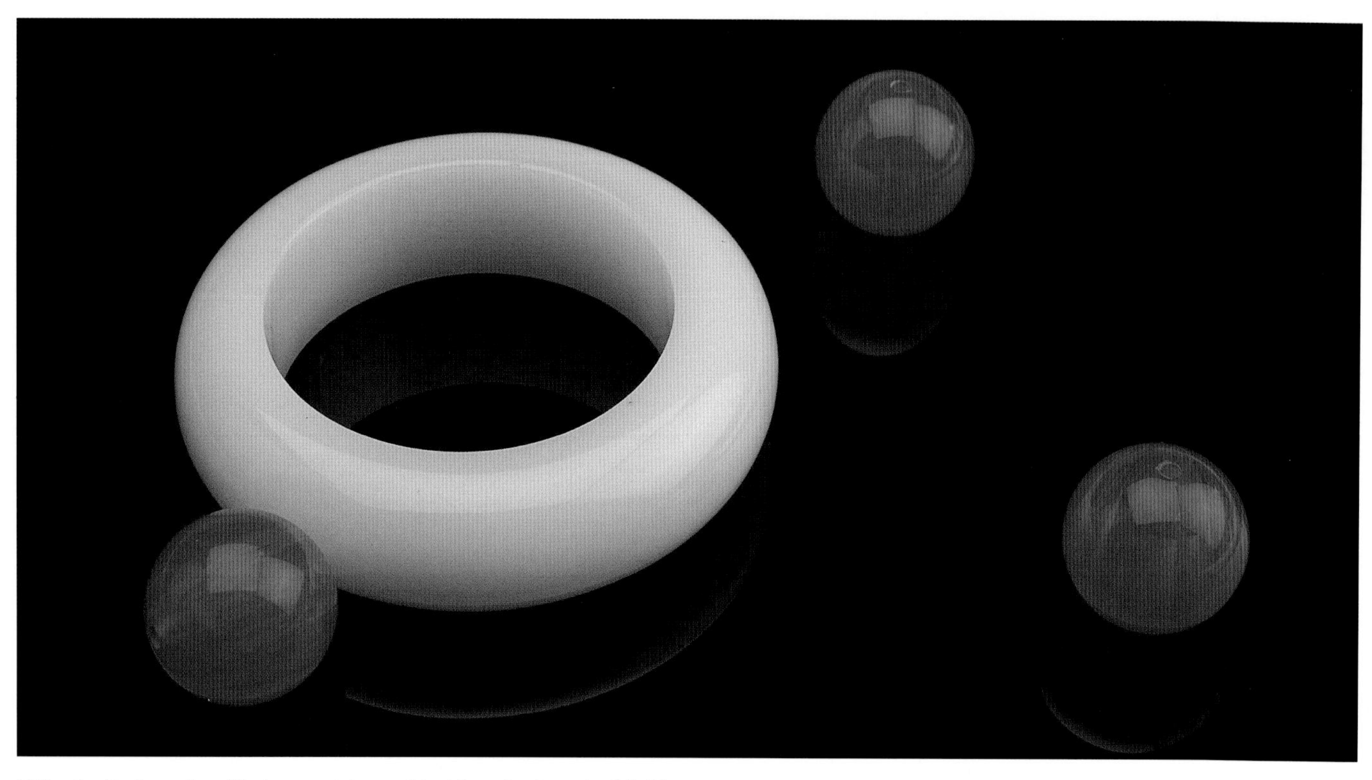

Milky lucite bangle with large pink marbled beads, bangle 25-30.

## Matches Made in Plastic Heaven

In this section, we have assembled some of our favorite groupings of bangles and other pieces, to suggest just a few of the myriad ways that you can incorporate plastic bangles into your life. We have mixed different types of plastics, as well as old and new. As you collect, you will assemble your own favorite groupings. Give your creativity free rein and enjoy your plastic!

Green Bakelite hitchhiking hand with metal chain and pink painted fingernails; $450 and up; heavily carved green bangles, $150-200 each; pink Bakelite spacer, $25-30.

Green Bakelite bangle, $45-55; Bakelite pin with celluloid bells on brass chains, $150.

Two Lucite bangles, clear and pale aqua, $25-45; Lucite palm tree, "Souvenir of Orlando, Florida," $45.

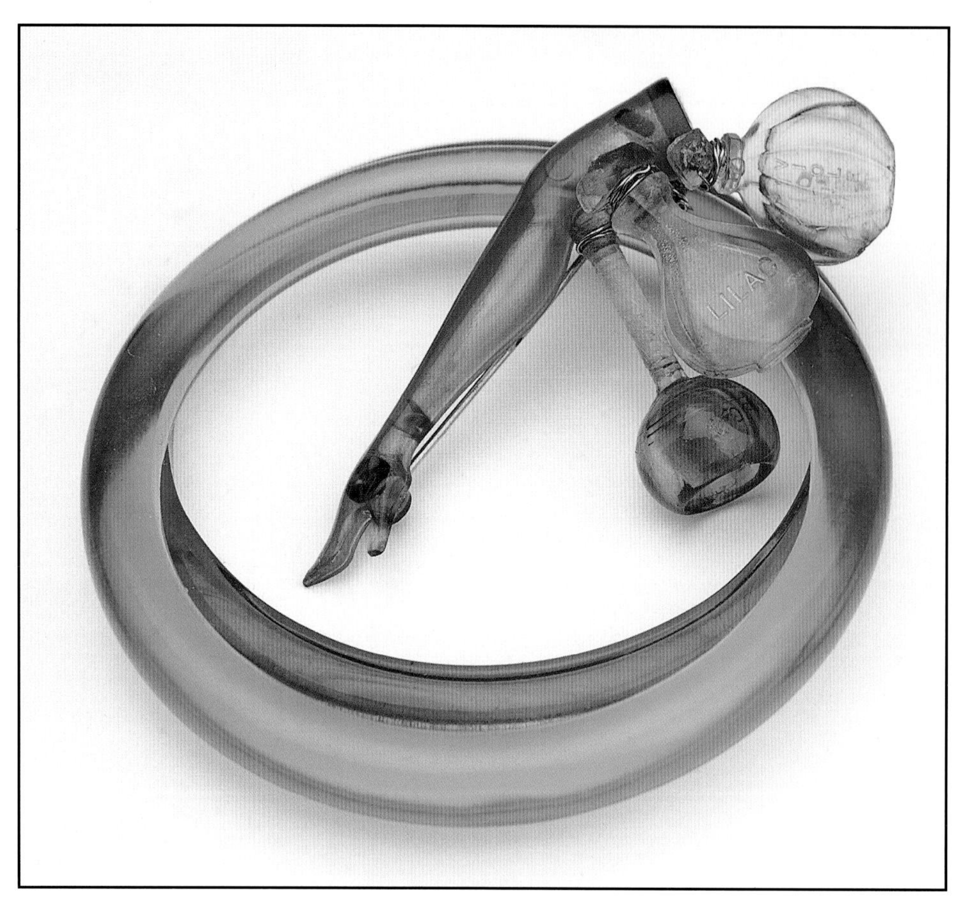


Plastic leg pin with perfume bottles, $40-45 with Lucite bangles, $5-8.

Lucite heart pin $25-35; two applejuice bangles, one Lucite and one bakelite, $10 and $25.

Vintage pearly bypass bangle, $35-45; Lucite clip earrings in black with pearlized inlay, by contemporary Italian designer Angela Caputi, $75.

An armload of moonglow and frosted carved Lucite bangles in matching and complementary patterns, $25-40.

Three moonglow Lucite or resin bangles. The top two are newer and probably American; the bottom one is a French vintage piece. $25-45.

Chapter 6

# A Visit to a Rhode Island Lucite Bangle Factory

Every collector's dream is to discover a factory where treasures have been sitting undisturbed for years. We were lucky enough to have this happen to us, and some of the results can be seen in the following pages. It was great fun to be able to buy "brand new" vintage bangles, but even better was the information we gathered about how they were made. Best of all, we were able to put a place and date to a large number of the pieces we had been collecting for years. It turned out, to our surprise, that something like 75% of the American vintage Lucite bangles we had collected came from one small factory!

The stock and equipment of Best Plastics of Providence, Rhode Island, was bought out by Plastic Development, Inc., and the operation relocated to Pawtucket, R.I around 1970. Lucite bangle bracelets and other matching pieces were made in first one and then the other location from the 1960s through the 1980s, although most of the multicolored bangles in the following pages date to the Best Plastics era (1960s). After the move, most of the bangles made were solid colored or marbled. Jewelry production tapered off and then ceased in the 1980s. Today Plastic Development no longer manufactures jewelry but continues to make other plastic items.

Factory in Pawtucket, Rhode Island, where thousands of Lucite bangles were produced in the 1970s, after Best Plastics was bought out by Plastic Development.

Extrusion-machine which produced the tubes to be sliced into bangles. A separate jet was used for each color in the design. There was a template for each color. All colors and layers were extruded through the machine together to form the tube, which was usually five to six feet long.

Grinding machine with grinding wheels. The bangles were sliced and shaped using centerless grinding. (The action is similar to that of a lathe.) The wheels shown were chosen according to the particular shape and size desired.

Polishing drum in which bangles were tumbled with wooden pegs or sawdust and polishing compound to achieve a shiny surface. An entire row of these drums was powered by a single motor.

An intrepid bangle excavator.

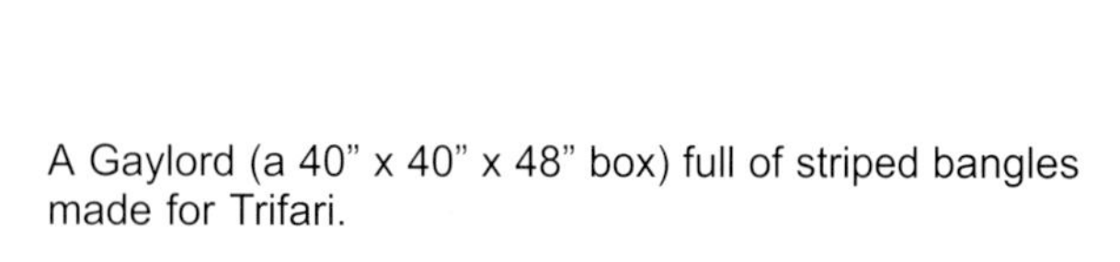

A Gaylord (a 40" x 40" x 48" box) full of striped bangles made for Trifari.

A group of striped Lucite bangles. A clear layer of Lucite over the color gives these the effect of cased glass. 1960s.

In its heyday, Best Plastics/Plastic Development made jewelry for a number of well-known American companies, such as Trifari, Napier, and Diane von Furstenberg. Other pieces have been located with tags for Tempo brand and for Sears. Located in the heart of one of America's largest jewelry producing centers, it nonetheless seems to have had little competition – or so we deduce from the fact that so many different types of bangles proved to originate from this one company. We discovered that the designs, which we had always imagined came from the jewelry companies, actually originated at the factory, which had its own designers. Styles were produced and offered to buyers at shows. Occasionally pieces were made to the specifications of the jewelry company but this seems to have been the exception. Best Plastics pieces were made in single colors, opaque, transparent, and marbled, but the most exciting pieces use at least three colors and take advantage of Lucite's unique refracting properties. Our favorites, which we fell into the habit of calling "cased," borrowing a term from glassblowing, are stripes and other simple patterns with a layer of clear Lucite over them. We learned that these layers were not added but were all extruded simultaneously to make the multi-colored tubes from which the bangles were then sliced. Each color had a separate jet, and each jet had its own template. These templates were reused in order to create the same stripes and swirls. As a result, you can take striped bangles of the same size and color and line them up, and in most cases the stripes will match exactly.

Lengths of sliced tubing went into a machine that cut them into the requisite number of bangles. The bangles were then placed in a polishing drum with wooden pegs or sawdust, and pumice, and tumbled for 24 hours, until they had achieved a smooth and shiny surface. Some more elaborate shapes required work on the lathe. We found a few unusual pieces that seem to be the result of individual experiments, including an interesting green and white carved bangle [seen on page 169 (bottomright) ].

Tone-on-tone striped cased Lucite bangles with matching earrings and findings. This type was sold in six colors: blue, gold, green, pink, coral-red, and aqua, with the last two colors being hard to find.

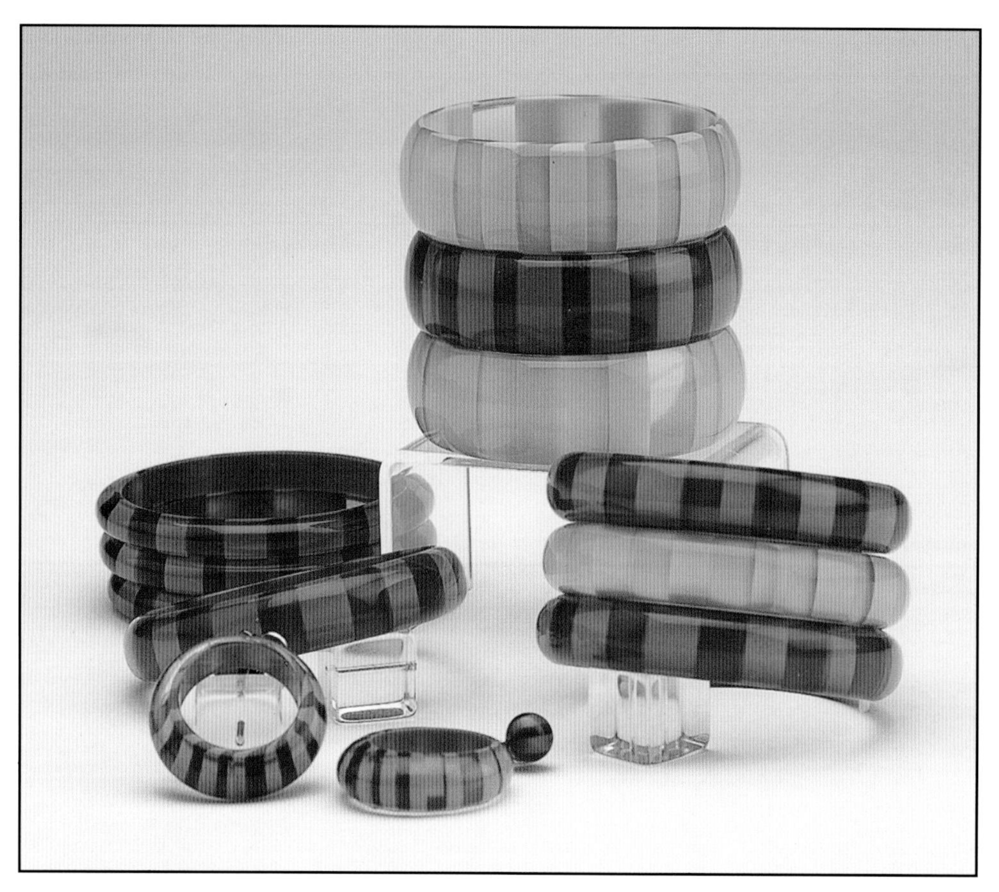

Pink and lime-green tone-on-tone striped cased Lucite bangles. Pink is fairly common but lime-green was apparently not sold. These are leftover stock from the factory. The 5/8$^{th}$ inch bangles to the right are also uncirculated stock, while the 7/8$^{th}$ and 3/8$^{th}$ sizes are easy to find.

A group of cased bangles, tone-on-tone aqua stripe, and an unusual yellow and aqua, $15-20.

Tone-on-tone striped cased Lucite bangles and matching earrings. The aqua flying saucer bangle is uncirculated factory stock. Bangles $5-20; earrings $10-20.

Opaque striped cased bangles in blues, greens, yellows, and black, 1960s, $6-45.

Pink and blue and pink and purple opaque striped cased Lucite bangles. Bangles $5-20, earrings $20.

Rootbeer and cream cased stripe bangles, earrings, and earring findings. Bangles $10-20, earrings $20.

Orange and yellow striped opaque cased Lucite bangles and matching pieces. Bangles $5-30; wide bangle $45; necklace $35; earrings $15-20. (Wide bangle and pierced earrings are very hard to find.)

A variant on the "cased" stripe, in two shades of green and in black and red, $7-15.

A black and white cased bangle, with a white striped with red and black bangle, probably made for Trifari, $12-15.

Fabulous chess pieces made from Lucite rods like the ones used for buttons and earrings. Best Plastics, 1960s. It is not clear how many of these sets were produced or sold.

White and brown cased striped Lucite bangles and earrings. Bangles $10-40; earrings $20-25.

Striped bangles with cased effect, 1960s. White with colors.

An unusually wide black and white striped cased Lucite bangle.

White and black or pink striped cased Lucite bangles and earrings. Bangles $10-25; wide black and white bangle, very hard to find, $50-60; earrings $25-30.

White cased with colored stripes. Bangles, $4-20; wide bangles (over one inch), $40 and up; earrings $15-20.

Cards showing findings for earrings and related pieces, from the Best Plastics factory.

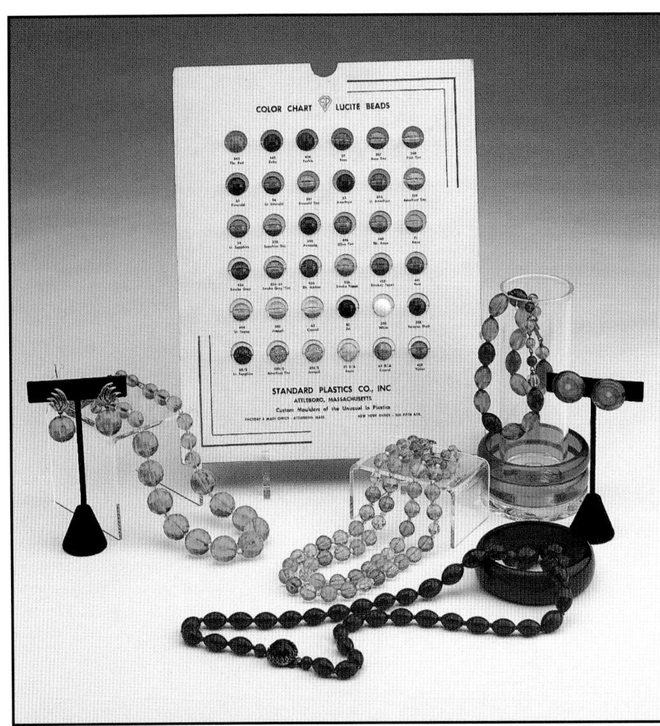


A sample board showing facetted Lucite beads from Standard Plastics, another plastic jewelry manufacturer in nearby Attleboro, Massachusetts, with necklaces made from similar beads.

Transparent striped and quadrant-patterned bangles in all three color combinations, with matching necklaces and earrings, 1960s. Necklaces $20-35.

Transparent striped narrow bangles, $8-10.

Transparent pink and yellow striped and quadrant bangles. Bangles $10-40; earings, $20.

Blue and green cased transparent striped and quadrant Lucites. Bangles $10-40; earrings $20 and up.

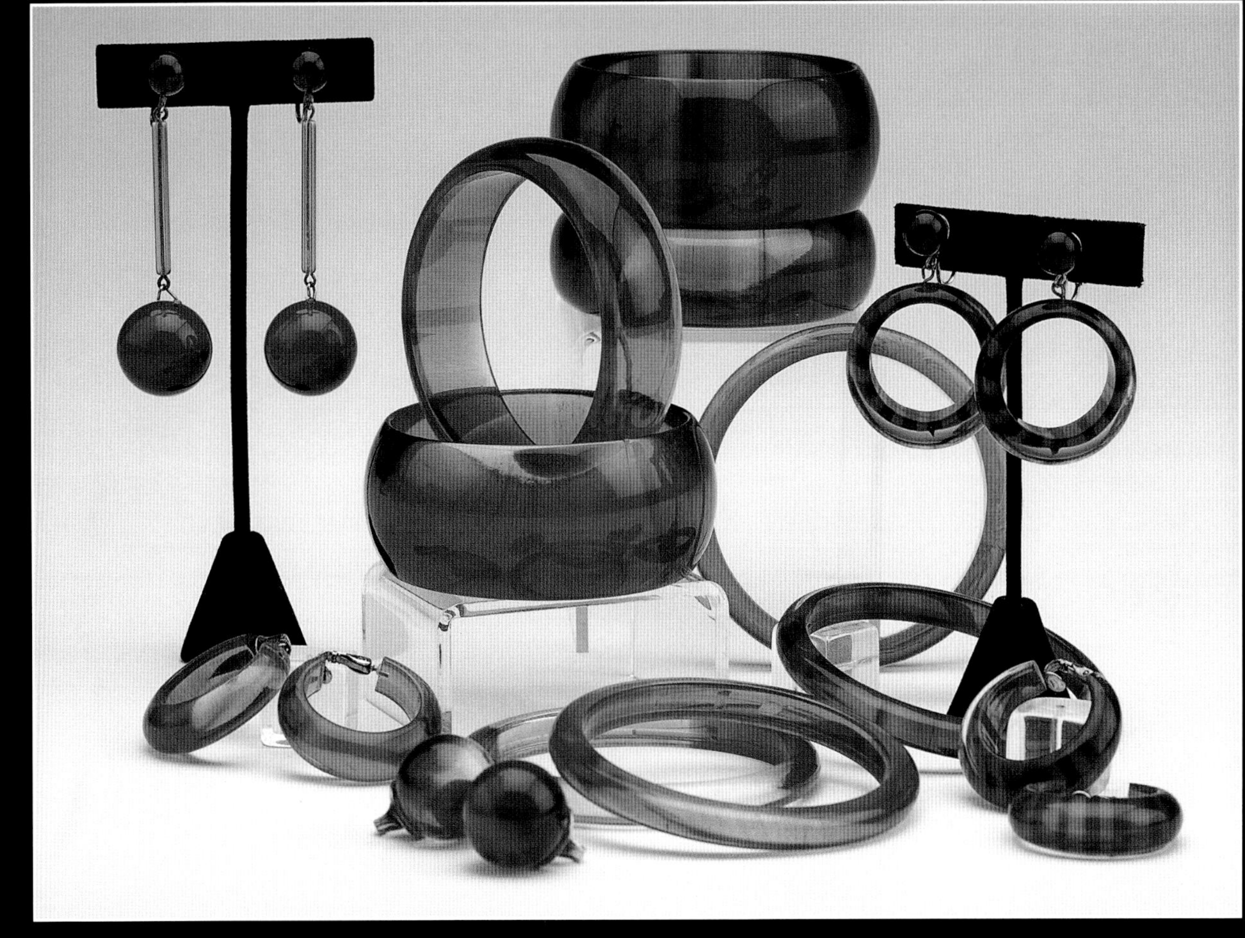

Pink and purple transparent striped and quadrant Lucites. Bangles $10-40; earrings $15 and up.

Transparent multi-colored striped Lucite bangles with matching earrings, 1960s.

Wide chunky transparent multi-colored bangles, sliced with rounded edges. These are all uncirculated factory stock, 1960s, $35-45.

Multi-layered transparent Lucite bangles with matching earrings, which give an idea of the layering effect of the bangles. The lavender and green ones have an inner layer of blue. Bangles $12-25; earrings $15-25.

Pair of layered translucent saucer bangles, $18-25.

Multi-layered bangles with two shades of the same color or two contrasting colors dividing the bangle. Bangles, $15-20; earrings $20.

View of layered bangle from the side showing the layers.

Domed version of the layered bangle, with striped transparent bangle in Saturn shape, $20-25 each.

Yellow transparent Lucite bangles with opposing areas of red and green, with matching pieces, 1960s. Bangles $10-40; earrings $15-20; necklaces 35-50; dangling cube earrings, $30.

Close-up of three watercolor patchy bangles, $12-15.

Pastel transparent Lucite with contrasting watercolor areas, 1960s. Lavender with blue and green is the most common of these. The others are uncirculated factory stock. Bangles, $5-20; earrings, $20; necklace, $35-40.

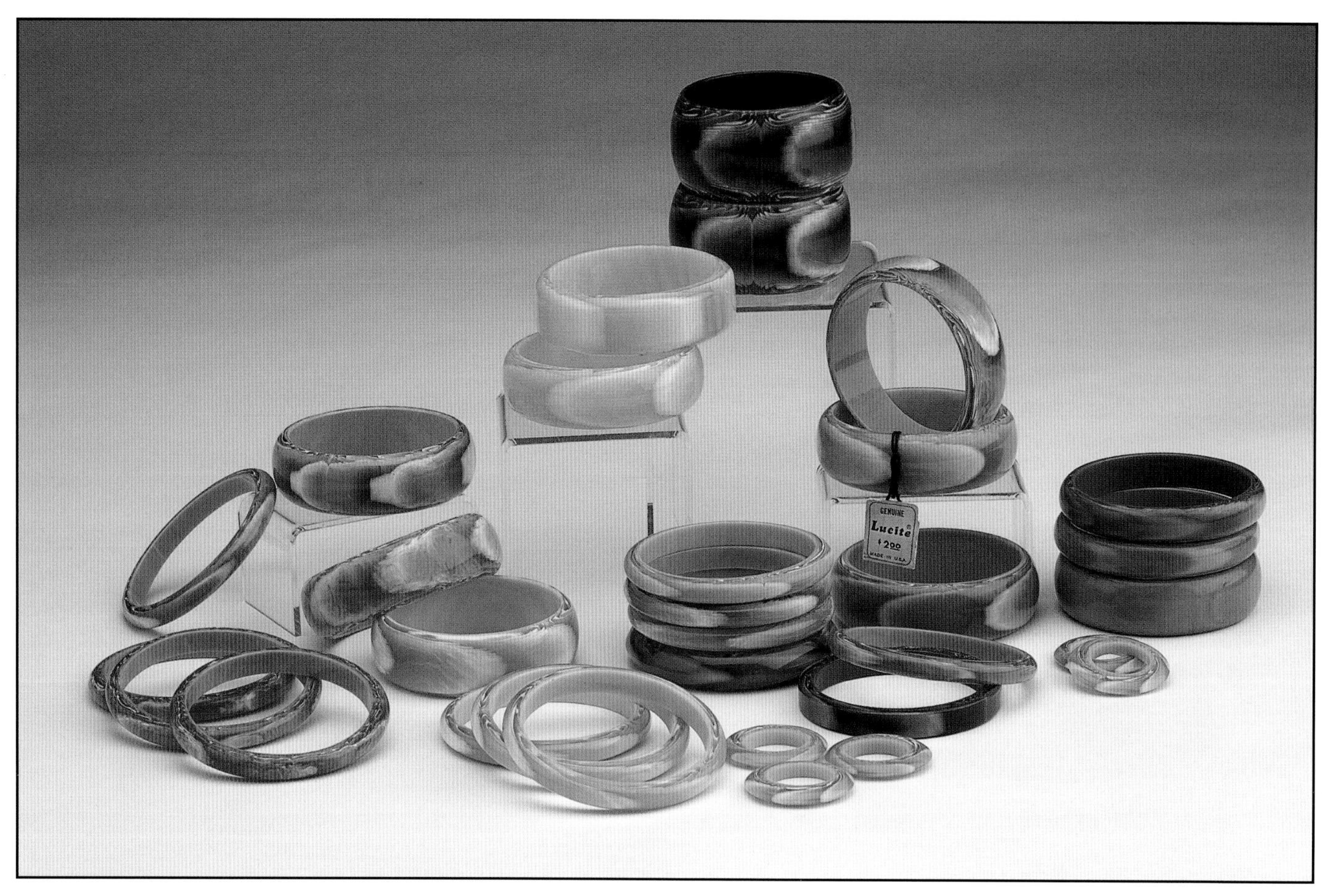


Moiré-patterned Lucite bangles, 1960s. Those on the right in multiple colors are probably experimental. The large size is hard to find, as is the deep blue color. $5-35.

Note how the beveled edge on the top yellow bangle creates a squared-off look in contrast to the wave-like pattern on the other bangles.

Moiré bangle with matching "buttons" probably intended for pins.

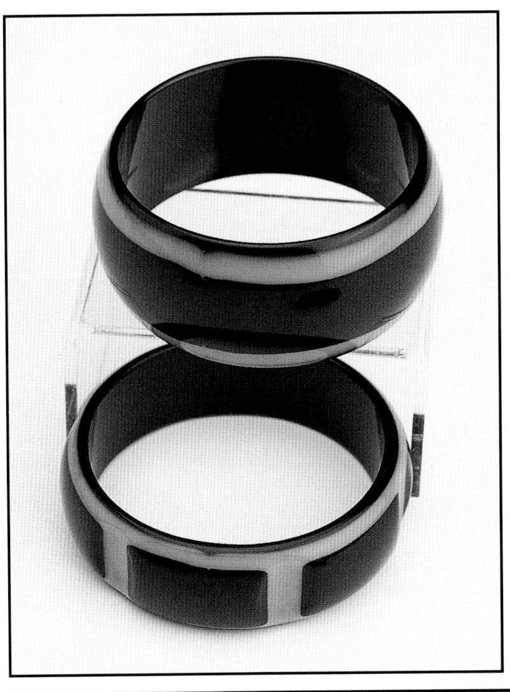

Black and yellow cut-back and matching striped bangle. This color combination as well as red with yellow appear as Bakelite in several books. Note the "indentation" in the yellow layer on each bangle. This is a clue to the extrusion process by which these were manufactured and proof that they are Lucite, not Bakelite.

Lucite layered bangles cut back in a "pillow" design with matching striped bangles and other pieces, 1960s. The green and white pillow bangle is a prototype or experiment. Pillow bangles $25-50; other bangles $10-25; necklaces $20; earrings $15; ring $25-35 (very hard to find).

Three patchy moonglow bangles, $10-18.

A rainbow of Lucite bangles with patches of moonglow or catseye pearlescence. Bangles $3-18; rings $5; necklaces $25.

Metallic moonglow bangles with contrasting colored patches, $15-20.

Opaque Lucite bangles with patches of moonglow color, 1960s. Most common are the white with darker and lighter shades of the same color. Silver or gold moonglow with iridescent patches are also found. Less common variants include red, white, and blue; black and white; and pastel pink or orange with moonglow patches of the same color. Single bangles and groups of very thin ones, $5-30; necklace $20-30.

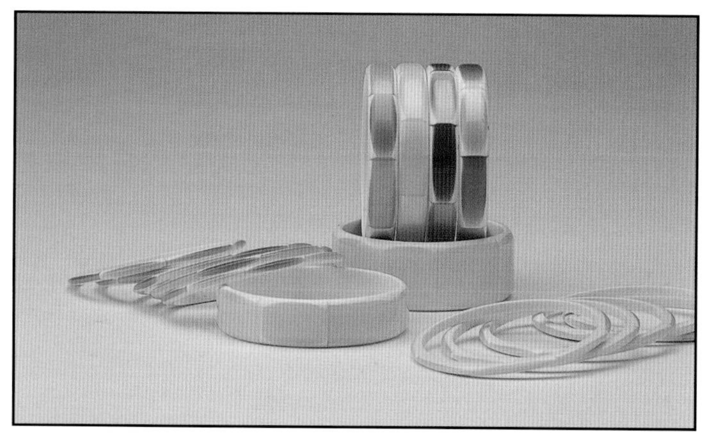

Opaque bangles with moonglow patches, wide, $30; medium, $20; narrow, $12; group of skinny bangles, $15.

White bangles with pinstripes, $15-35.

Opaque, translucent, and transparent bangles with narrow "pinstripes," 1960s, $8-35.

Two tortoise with green pinstripe bangles, narrow, $10-15; wide, $30-35.

Opaque Lucite bangles with stripes, either contrasting colors or two shades of the same color, shown with findings from factory, 1960s. Single bangles or groups of narrow ones, $5-35; earrings $10; on original card, $15.

Opaque striped bangles in pink and blue with matching earrings. Bangles, $9-12; earrings, $10-15.

Wide cream bangle with blue-green and green stripes, $35-40.

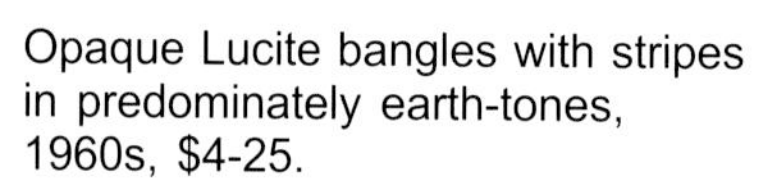

Opaque Lucite bangles with stripes in predominately earth-tones, 1960s, $4-25.

Translucent and opaque Lucite bangles each with four sections of white or cream, 1960s, $5-40.

Two groups of Lucite bangles, one with irregular black patches on bright colors, and other with swirls of black and cream on muted colors, 1960s. Patches $10; swirls $10-25.

Opaque Lucite bangles ringed with triangles in two shades or two colors, 1960s. Bangles $5-20; earrings $10.

Wide Lucite bangles in a range of colors with raised black rings, also all white, $10-12.

Tortoise-shell Lucite bangles with opaque stripes of white, black, tan, blue, and orange, 1960s. Single bangles or groups of narrow ones, $5-35.

Two sets of narrow bangles, 1960s, with Trifari labels, originally sold for $3. Current value, $25; $15.

Multi-layered Lucite bangles, 1970s. The purple and pink ones have a transparent under-layer and an opaque layer of pink. $10-15.

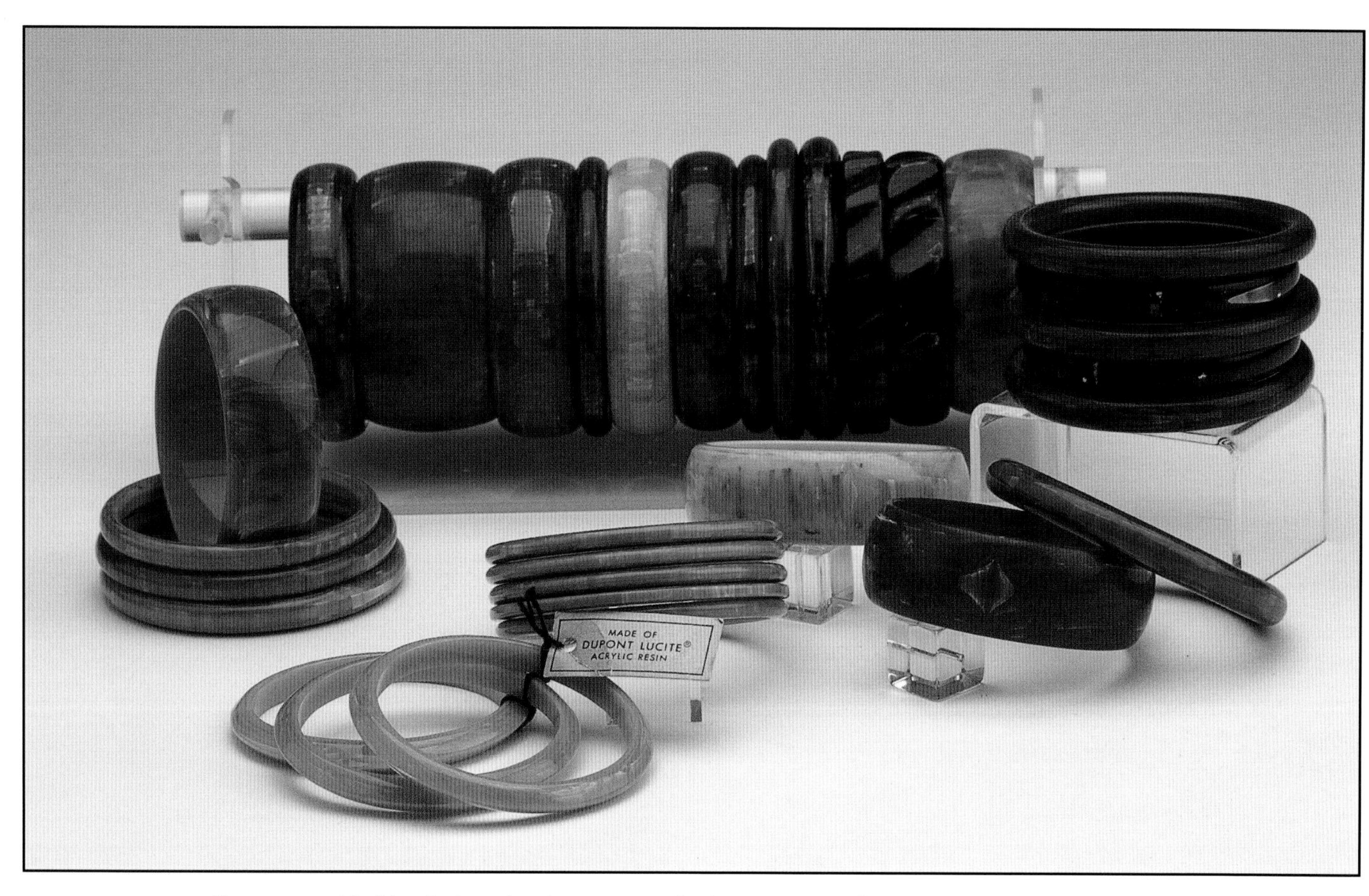


Opaque marbled Lucite bangles, in a range of colors, probably from the 1970s. Sometimes mistaken for Bakelite, they can be identified by the whiteness of the marbling. Note the label reading "Made of Dupont Lucite® Acrylic Resin" on the set of three orange spacers. $5-20.

Group of muted marble bangles and oval-shaped bangles in solid colors, 1970s, $15-25.

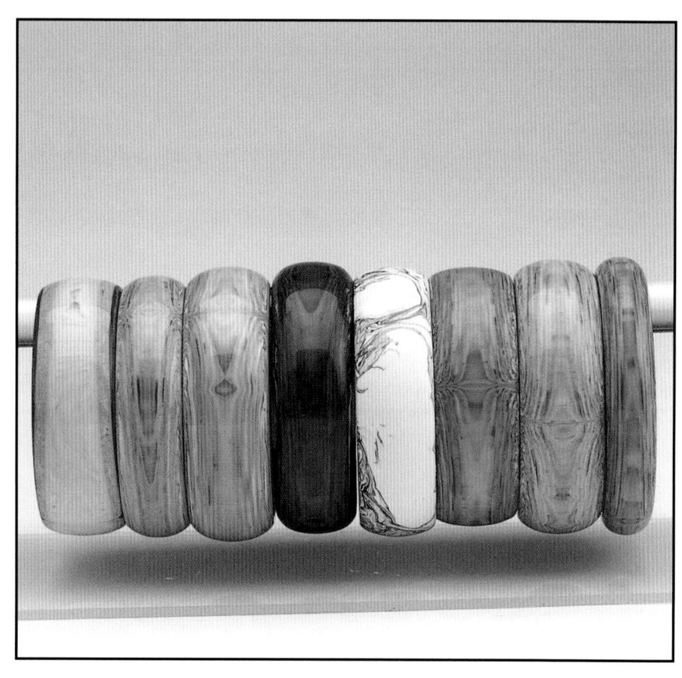

One-inch marbled bangles, $8-12.

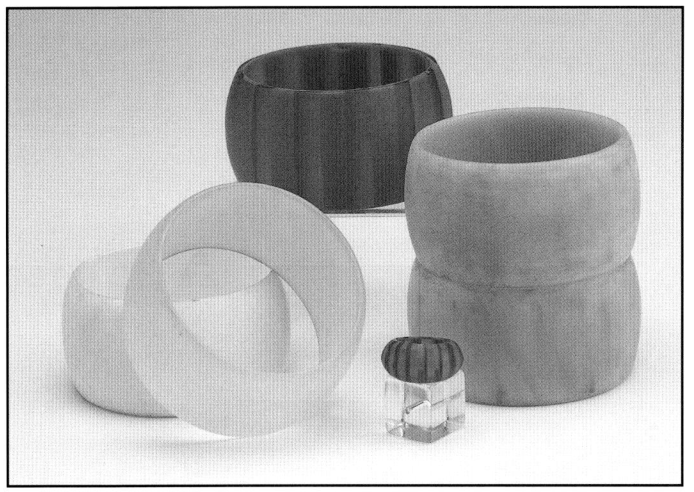

Wide frosted bangles in marbled pastels with an unfinished set of blue tone-on-tone bangle and ring. Pastels, $15-20; blue set, very hard to find, $35-45.

Satin-finish moonglow bangles, narrow, $10; wide, 25.

A riot of opaque Lucite bangles with colored swirls, 1960s. Most have two shades of the same color on white or cream, but note grey with black and pink, and white with tan and dark brown, among other variants. Single bangles and groups of narrowest bangles $4-20; bangles over one inch, $30-40; rings $5.

Stack of three swirl bangles, $10-15.

Three sliced swirl bangles, $10-15.

Beveled and domed swirl bangles, $10-40.

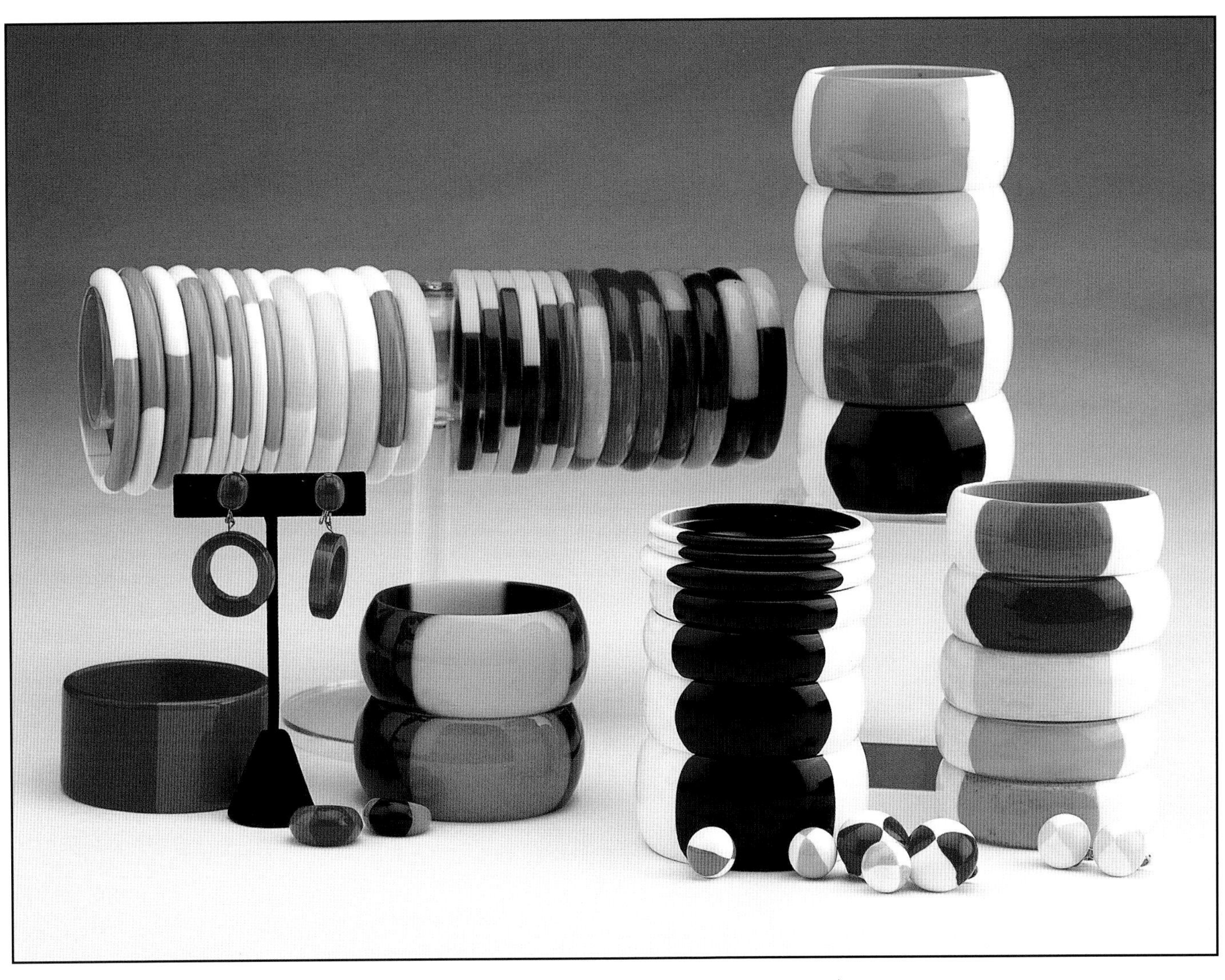

Lucite quadrants of opaque color alternating with white or tan, 1960s, $10-35. Those with two shades of the same color are harder to find. Bangles, $12-45; rings and earrings, $10-25.

Lucite bangles with two shades of red or green, frequently mistaken for Bakelite. Set of five narrow sliced, $25; narrow, $10; wide, $35-45.

Narrow bangles with two colors alternating in quadrants, 1960s, $10-15.

These patterns, usually made in one color on white, were also made in patriotic red, white, and blue versions, 1960s, $10-20.

An explosion of red, white, and blue, many manufactured by Best Plastics. (Best Plastics: Second stack from left, two sliced bangles in left stack, dome and dangling earrings with "quadrant" pattern, twist bracelet at the top with "quadrant" balls at the ends, and the matching uncut rod, center.) The laminated striped bangles were made elsewhere.

Chapter 7

# Shopping for Vintage Plastic Bangles

## Where to Find Vintage Plastic Bangles

This is a relatively new collecting field, so you will need to be resourceful. That is in fact a great part of the fun – plastic bangles can show up just about anywhere.

Start with the obvious – flea markets, antique shows, yard sales, and estate sales. You will need some patience, which will be repaid amply when you score that great find. Because collectible bangles are not always recognized as valuable, you can find them in thrift shops like the Salvation Army or Goodwill, if you have sharp eyes. Consignment and vintage clothing shops are also a good bet. Be aware however that antique dealers sometimes ignore all but the most spectacular pieces, while dealers in collectibles sometimes overprice rather modest examples. Talk to dealers and let them know that there is a market for reasonably priced plastic jewelry. Get them to call you when new things come in. Some dealers avoid plastics because they are uncertain about how to identify them, and may be interested in learning more. Time spent developing these relationships can be very worthwhile. If you are lucky enough to have the chance to explore the flea markets of London and Paris, you will find things that rarely turn up in the United States.

It is no longer necessary to hit the road in order to find good things. Many dealers now have on-line stores, either through Ebay or through specific hosts for antique dealers. You can find these by Googling "Lucite bangles" or whatever it is you are looking for. Be sure to experiment with different ways to word your search. Dealers' websites can tide you over during the winter months, when flea markets are scarce. Be aware, however, that prices on these websites vary greatly from cheap to outrageous, and not all dealers are familiar with the plastics they sell.

On-line auctions are another very important source of vintage plastic. The allure is obvious – hundreds of new pieces come on the market everyday. But here the old phrase *caveat emptor* ("let the buyer beware!") could not be more relevant. Recently scores of questionable pieces have flooded the market. Be sure to exercise caution and skepticism when bidding. Armed with knowledge, you can find great pieces at great prices. Ask questions, ask for more photos, and check the pieces against the ones in this book. Before taking a chance on an iffy piece, check the seller's return policy. Don't get carried away by the excitement of the auction. Most of the pieces you will be bidding on were made by the hundreds or thousands. If something you want has gone too high, don't get into a bidding war. Let it go. Another one will come around sooner or later. And don't let wishful thinking cloud your judgment. That intriguing piece in the back could be something rare and special, but it might also be junk! Check out not only the seller's feedback, but also other pieces a seller has up or has sold in the past. This is important because buyers may not realize immediately that they have been sold new rather than vintage jewelry. Some well-intentioned dealers may not be aware of the origin of the pieces they offer for sale, and are grateful for information about them. Be especially cautious about bidding if the seller has private feedback or if the bidders' list is kept private. Caveats aside, we have had good experiences with electronic auctions. On-line bidding has put us in touch with many other delightful and knowledgeable sellers and collectors. We even met each other by bidding on the same bangle!

## What to Look for When Shopping

Anything that appeals to you is worth collecting. We recommend that you look for colors that call out to you, elegant or whimsical designs, and good condition. Substantial, well-made pieces will give you more pleasure and appreciate in value more than flimsy ones. We have included a wide range of types at different levels of quality in this book, to let you know what's out there. Bangles that appear in books are destined to be called "book pieces," but in this collecting category, that is not enough to make them valuable. Some of these are worth a great deal, and some are worth very little – unless you happen to love them!

You may choose to collect at random or to assemble as many different colors of the same type as you can. You may want to wear everything you own or reserve it for display. One of us particularly likes to find all the matching earrings, necklaces, and other related pieces to go with our favorite bangles. The other concentrates on assembling stacks of bangles that complement each other. You might decide to specialize in celluloid, or West German bangles, baby bangles, or laminates. It's all up to you – and your budget!

# Condition

Condition is important, as all collectors know. You may occasionally decide to acquire a less than perfect bangle inexpensively just to wear and enjoy, but if you should decide to sell it, expect to get your money back at best.

Things you need to look out for when considering a purchase:

Beyond the obvious – cracks, chips, splits, scratches, and pits – some other problems that beset plastics are fogging, especially noticeable in clear or pearly Lucite; warping; and fine crackling inside clear pieces. When buying any piece with hardware, check for missing parts. If it's a hinge bracelet, are all the screws in place and does the hinge snap shut and stay shut? (A little play is not a problem but you want the thing to stay on your arm.) Memory-wire bracelets usually have free-hanging beads at either end. The last bead should be finished off with a flat end, like the head of a pin, rather than a loop, which would indicate that something is missing. Beaded necklaces can be missing a strand or some of the beads at the end of an extender. Plastic necklaces often have a metal clasp with a plastic piece (usually half a bead) glued on, so make sure that piece is still in place. On hinge bracelets with a metal frame, check for empty loops betraying the loss of a safety chain.

Celluloid can deteriorate and become "sick," a sign that it is on its way to disintegration. Sick pieces can "infect" healthy ones, so be sure to segregate any piece that looks questionable and of course avoid acquiring pieces that have begun to degrade.

Prystal (left) and hard plastic (right) bangles with internal damage. Note the obvious cracks running through the Prystal, while the other bangle has a network of fine crazing.

Identical Monet frosted Lucite bangles, one with signature plaque, the other showing an empty indentation where the plaque should be. Intact bangle, $35-45; damaged bangle, $10-15.

Blue moonglow memory-wire bracelet, missing one end bead.

## Care and Repair of Vintage Plastic Jewelry

Lucite jewelry is not terribly fragile, but should not be thrown together to rattle around in a box or package, as scratching can occur. Older and more delicate pieces obviously require more care. Bakelite should be stored out of the sun, as it will darken the colors. (Prystal is especially prone to cracking if exposed to extremes of temperature, either hot or cold.) If you live in a northern clime, avoid wearing Bakelite outside in the winter. Thermoplastics, especially the softer ones, can become distorted if exposed to heat for too long. Celluloid needs to breathe. Never store it in a plastic bag, since both the celluloid and the bag give off gasses that can interact. You could then find yourself with a bag full of nothing but crumbs and a pungent odor. (See above on "sick" celluloid.) Galalith (like celluloid) should not be stored in hot showcases where the air is too dry, or be subjected to sudden temperature shifts.

Don't be afraid to clean your plastics, but use common sense. Wiping with a damp cloth with a little bit of mild soap if needed, is fine for most pieces, but take care not to damage any decorations. Bakelite, Lucite, and hard plastic can be soaked in water and scrubbed with a toothbrush to get dirt out of crevices. Celluloid should not be left in water for any length of time, as it can swell and become distorted. Painted pieces should be cleaned with care, testing on an inconspicuous place to make sure the paint is stable. Bakelite can be cleaned with solvents if necessary, to remove undesirable or damaged paint, but the surface polish will need to be laboriously restored afterwards. Avoid using any kind of solvent in the vicinity of any of the other plastics, as they can easily be ruined. Rhinestones require extra care. Be sure to dry carefully using a hairdryer, because moisture trapped under the stones can cause them to become dull or leave visible rust in the settings.

Bakelite, Lucite, and other hard-surfaced plastics can be polished with Simichrome or Turtle Wax and buffed with a soft cloth. Novus products can be helpful in removing scratches and polishing. Do not attempt to use these on celluloid, or on any piece that may have been tinted or painted.

Minor repairs can be made using epoxy or Crazy Glue, which now comes in a bottle with a brush. This allows you to get the glue exactly where you want it to go. Epoxy has the stronger hold but is a bit harder to work with. Valuable pieces and more difficult repairs should be taken to a jeweler. We occasionally hear rumors of people specializing in the repair of Bakelite, but they are hard to track down. Bakelite artists will sometimes agree to repair vintage pieces, but most are too busy making new ones. Keep this in mind when considering buying a piece that needs work.

Chapter 8

# New Plastics

## Cheap Imposters

"Cheap and cheerful" describes these fun retro Lucite and hard plastic pieces from Asia. Especially popular are bright colored bangles with large white dots, and layered laminates. These are, on the whole, sturdy and well-made. In addition to these, department stores have started carrying reverse-painted hard plastic bangles. While the colors are fun, the quality is poor. Many of them don't even get out of the store without breakage, in marked contrast to the many vintage pieces still to be found in good condition after forty or fifty years. Another group is frosted and reverse-painted to resemble expensive contemporary pieces like those of Alexis Bittar (see below). Here the inside finish is the giveaway that these are cheaply made pieces. From the outside they are rather attractive, although without the glow that characterizes Bittar's work. Our view – if you enjoy these imposters, why not buy a few, on sale of course! Just don't fall for the sellers who claim that they are selling "French Bakelite."

A fun layered Lucite bangle from China, $10 retail. Sometimes compared to the work of Lea Stein, this design is much chunkier than anything she has produced.

A sampling of new plastic pieces made in China of Lucite or another modern plastic. Most of these retail for between $5 and $15 in shops and flea markets in the U.S., London, and Paris. They have been seen on Ebay and elsewhere offered at much higher prices as "French Bakelite" or Galalith, with the suggestion that they are vintage.

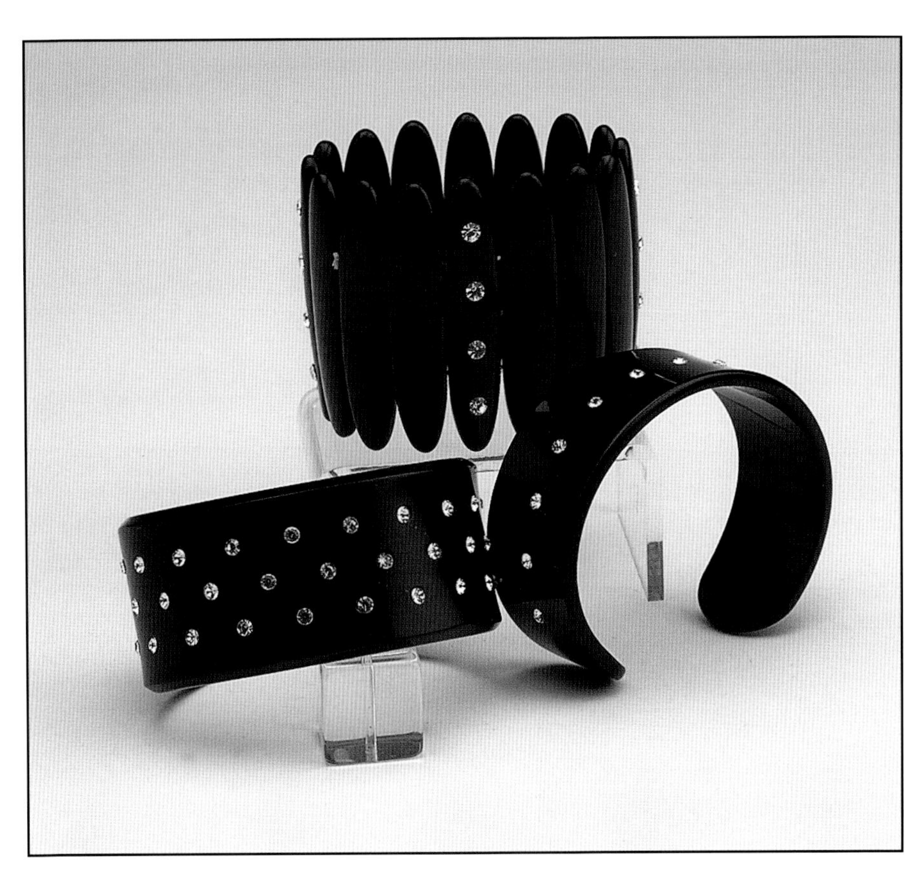

Black cuffs and stretchy bracelet made in China, $7-12 retail.

A pleasingly weighty freeform bangle with rainbow coloring, probably made in China, also found in solid transparent colors, $15-20 retail.

# Expensive Imposters

We are not kidding when we say expensive. Many a collector has been chagrined to discover that she has spent good money to acquire pieces made last week in Taiwan, China, or even in the U.S. These fall into two categories.

A newly carved bakelite hinge bracelet, open to show the round shape. Vintage hinge bracelets are usually oval, not round, in shape. This pattern is frequently seen in newly carved bangles, usually black. $45.

### Newly Carved

For quite some time, pieces have been circulating that look and feel like vintage Bakelite, except that something about the carving is off. Once you recognize the few patterns that turn up over and over again, you will not be taken in. Characteristic of these pieces is a lack of subtlety in the carving, which consists in part of rather wide shallow cuts. The palette is also quite limited. Solid yellow, red, black, and grayish-blue are the usual suspects here. As far as we can tell, these pieces are usually made from old Bakelite stock, but some may be carved from other materials. The prices on these pieces tend to be high because they are mistaken for vintage by novice collectors.

Newly carved Bakelite bangles, probably from old stock. Each of these passed the smell and Simichrome tests. The yellow bangle shows a particularly common design, found in different colors and widths. The patterns attempt to imitate those of vintage Bakelite, but will not fool the experienced collector. Note the lack of subtlety in the carving. $20-45.

The same hinge bracelet closed.

### Fakelite

This term has been used for a number of different kinds of deception, including the assembling of old buttons and buckles to make new pins, but recently it has been reserved for the avalanche of newly made pieces from Asia. Over time these imposters have become more convincing. This is a serious problem affecting Bakelite values across the board, especially on Ebay. Now that word is out about the fakes, dealers and collectors alike are hesitant to buy. The result is that while the most sought-after pieces (figurals, random dot bangles, and other rarities) still fetch high prices, good pieces in the middle range, especially heavily carved pieces, are simply not showing up on the market as often. Their owners are no doubt holding back until things settle down and they can be assured of getting a good price again.

The best defense against these fakes is to study the many good Bakelite books out there (listed in "Further Reading") until you develop a good instinct for what vintage patterns and styles look like. All of these books were published before the advent of Asian fakelite. Most of the fakes are hefty, wide, heavily carved bangles, but pins are also turning up. Somewhat fleshy leaves and flowers are popular with the carvers, as well as improbable animals like octopuses, elephants, and geckos. Recently designs with lips have also been appearing. Dogs and especially cats should also set off your fakelite detector. Some of these bangles have been carved away along the edges, until all that remains is a chain of geckos or other animals. While there is always a chance that an unknown vintage design will surface, your suspicions should be aroused by anyone who has several of these. Alas, it seems that some of the producers have begun referring to collectors' books for ideas, and have started producing copies of such rarities as the famous "Andy Warhol" willow tree bangle (in Davidov and Dawes).

Be suspicious of any seller who introduces such a piece too casually or has tons of unusual heavy carved bangles. Aside from a few well-known dealers who only handle very high-end pieces, most sellers will only come upon a few of these showpieces in a lifetime. You should expect to see a high reserve and some excitement on the part of the seller if the piece is genuine. The seller may say that the piece passes Bakelite tests, without actually saying that it is vintage. Any questions you ask about the item should be answered straightforwardly. Private feedback and private auctions should also set off alarm bells. Needless to say, any reputable dealer will allow you to return a piece if you are not satisfied that it is vintage.

Other clues: colors are rarely consistent throughout the piece, and are sometimes completely inconsistent with Bakelite manufacturing techniques. A spot of unrelated color in the middle of an otherwise solid or marbled piece is a tipoff. Some fakelite is attractive in color, but much of it is muddy, and even ugly. Fakelite bangles are somewhat heavier than Bakelite, and often have a slightly rounded inner lip, rather than the straight up and down edge of true Bakelite. Finally, although some of these pieces are at least partly hand-carved, others have been molded to look as if they were. See "Testing" for further information on how to detect fakelite. Although some fakelite has been doctored to "pass" Bakelite tests, the results are in fact not the same as for the genuine article.

Asian fakelite bangles compared to vintage Bakelite. Top, vintage Bakelite on left; middle, vintage cast carved Bakelite bangle on left; bottom, both are fakes. Translucent fakelite tends to be darker around the edges, while real Bakelite is usually lighter around the edges. Note the distribution of the two colors on the orange and black bangle. The uneven rings of color are characteristic of fakelike, but are never found in real Bakelite.

Close-up of Bakelite (left) and fakelite (right) bangles. Compare the depth and delicacy of the carving on the Bakelite. The fakelite may in fact be molded.

This bangle is an excellent (and cautionary) example of all that is wrong with fakelite: the unpleasantly brassy yellow color, the transition to green about a third of the way down, and the oddly fleshy petals and leaves. The extreme chunkiness and awkward proportions of this bangle are also typical. Courtesy Lisa Sachs.

# Designer Pieces – Tomorrow's Collectibles

**Alexis Bittar**, a New York artist and designer, works in Lucite. His bangles are made of clear Lucite coated on the inside with color and gold paint or gold leaf, giving them an intense luminosity. The surfaces are hand-carved or textured and have a frosted appearance. His earlier work was monochromatic, but he has recently begun designing bangles and other pieces with multicolored designs. Made famous by the television show *Sex in the City* and a recent Estée Lauder ad campaign, his work always seems to be in short supply.

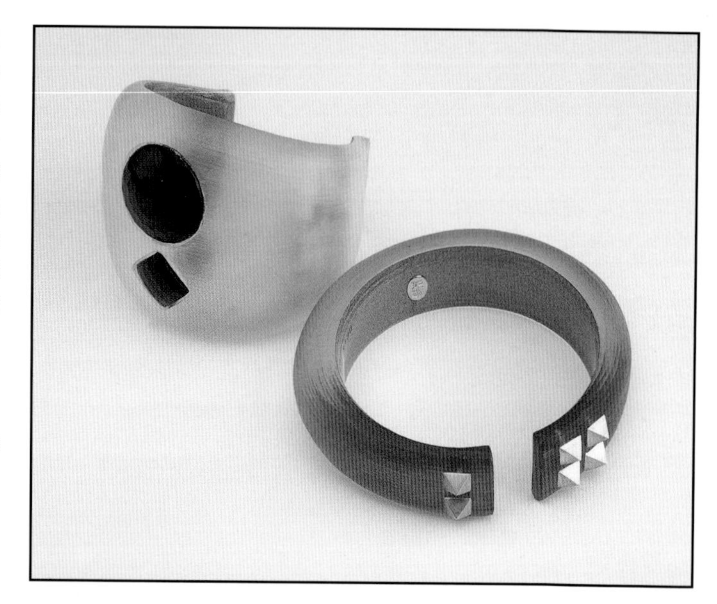

Alexis Bittar white cuff with black facetted cabochons, $125; magenta open bangle with gold studs, $130.

Alexis Bittar bangles, cuffs, and earrings. Bangles, plain narrow, $45-60; wider, $100; wide patterned bangles, $200 and up. Most pieces, courtesy Lori Kizer.

**Ron and Ester Shultz**, the first and most famous of contemporary Bakelite artists, began by imitating vintage pieces, but soon developed their own distinctive style. Their bangles feature interesting color combinations, often with dots or checks. All but their earliest pieces are signed. They are best known for laminates and inlays, but have also produced some elegant figural carved bangles. They invented the checked laminate bangle. Other well-known Bakelite artists whose work fetches top dollar are the Panttis and Howard and Karen Kronimus. All of these artists sign their work, although some early Shultz is not signed.

A fabulous red marbled signed Shultz Bakelite bangle with inlaid elongated oval dots, and an early signed Shultz bluemoon dot on cream bangle shown between two vintage bluemoon bangles. All Shultz pieces are made from vintage Bakelite, and all but the very earliest are signed. Red and black, $400 and up; bluemoon dots, $150 and up.

**Judith Evans**, an artist working in England, uses a phenolic compound without the formaldehyde that makes working with Bakelite hazardous. This compound also does not discolor over time. Her sense of color and proportion make her bangles and pins especially attractive. Checks and gumdrop bangles in colors that no one could mistake for vintage Bakelite are her specialty. Her pieces are signed.

An array of Judith Evans bangles, made of phenolic resin without formaldehyde. It looks and feels like Bakelite, but will not change color. All pieces are signed. Checks, $250-350; gumdrops, $185-250. Courtesy Lori Kizer.

Two signed Judith Evans bangles, a cream check with transparent pink, peach, and apple juice and a spectacular gumdrop bangle with the colors arranged in a spectrum. $230-250. Courtesy Lori Kizer.

**Rafia & Bossa**, an Italian husband-and-wife team working in Paris design these interesting bangles, which are made of acrylic resin. Their pieces are sometimes described as "French bakelite" (galalith) but this is inaccurate, as they themselves have confirmed. We particularly like their designs featuring translucent dots against a black background. Other styles feature geometric patterns. They bring out two new collections a year, loosely based on vintage designs, but in unusual colors.

This is just a small selection of the interesting work being done in various plastics at the moment. Keep an eye out and you may discover your own plastic collectibles of the future.

Rafia & Bossa bangles, made in France of Lucite or another acrylic resin (not "French Bakelite"), $55-65. All Rafia & Bossa bangles are signed on a small rectangular goldtone plaque. Courtesy Ann Barbour.

# Suggestions for Further Reading


Baker, Lillian. *Plastic Jewelry of the Twentieth Century: Identification and Value Guide*. Padukah, KY: Collector Books, 2003.

Battle, Dee and Alayne Lesser. *The Best of Bakelite and Other Plastic Jewelry*. Atglen, Pennsylvania: Schiffer Publishing Ltd., 1997.

Burkholz, Matthew L. *The Bakelite Collection*. Atglen, Pennsylvania: Schiffer Publishing Ltd., 1997.

Davidov, Corrine and Ginnie R. Dawes. *The Bakelite Jewelry Book*. New York: Abbeville, 1988.

Dinoto, Andrea. *Art Plastic: Designed for Living*. New York: Abbeville, 1984.

Ettinger, Roseann. *Forties and Fifties Popular Jewelry with Price Guide*. Atglen, Pennsylvania: Schiffer Publishing Ltd., 1994.

——. *Popular Jewelry of the 60s, 70s, and 80s*. Atglen, Pennsylvania: Schiffer Publishing Ltd.,1997.

Kelly, Lyngerda and Nancy Schiffer. *Plastic Jewelry*. Atglen, Pennsylvania: Schiffer Publishing Ltd., 2001.

Lauer, Keith and Julie Robinson. *Celluloid: Collectors' Reference and Value Guide*. Paducah, KY: Collector Books, 1999.

Meikle, Jeffrey L. *American Plastic: A Cultural History*. New Brunswick, NJ: Rutgers, 1997.

Parry, Karima. *Bakelite Bangles: Price and Identification Guide*. Iola, WI: Krause, 1999
.

———. *Shultz Bakelite Jewelry*, Atglen, Pennsylvania: Schiffer Publishing Ltd., 2002.

Razazadeh, Fred. *Costume Jewelry: A Practical Handbook*. Paducah, KY: Collector Books, 1997.

Wasserstrom, Donna and Leslie Pina. *Bakelite Jewelry: Good, Better, Best*. Atglen, Pennsylvania: Schiffer Publishing Ltd., 1997.